Mobility and IoT for the Smart Cities

Mobility and IoT for the Smart Cities

Editors

Luis Hernández-Callejo
Sergio Nesmachnow

MDPI • Basel • Beijing • Wuhan • Barcelona • Belgrade • Manchester • Tokyo • Cluj • Tianjin

Editors
Luis Hernández-Callejo
Universidad de Valladolid
Spain

Sergio Nesmachnow
Universidad de la República
Uruguay

Editorial Office
MDPI
St. Alban-Anlage 66
4052 Basel, Switzerland

This is a reprint of articles from the Special Issue published online in the open access journal *Smart Cities* (ISSN 2624-6511) (available at: https://www.mdpi.com/journal/smartcities/special_issues/Mobility_IoT).

For citation purposes, cite each article independently as indicated on the article page online and as indicated below:

LastName, A.A.; LastName, B.B.; LastName, C.C. Article Title. *Journal Name* **Year**, *Article Number*, Page Range.

ISBN 978-3-03943-050-5 (Hbk)
ISBN 978-3-03943-051-2 (PDF)

Contents

About the Editors vii

Preface to "Mobility and IoT for the Smart Cities" ix

Sergio Nesmachnow and Luis Hernández-Callejo
CITIES: Ibero-American Research Network for Sustainable, Efficient, and Integrated Smart Cities
Reprinted from: *Smart Cities* **2020**, *3*, 758–766, doi:10.3390/smartcities3030038 1

Marcos Lafoz, Gustavo Navarro, Jorge Torres, Álvaro Santiago, Jorge Nájera, Miguel Santos-Herran and Marcos Blanco
Power Supply Solution for Ultrahigh Speed Hyperloop Trains
Reprinted from: *Smart Cities* **2020**, *3*, 642–656, doi:10.3390/smartcities3030033 11

Silvina Hipogrosso and Sergio Nesmachnow
Analysis of Sustainable Public Transportation and Mobility Recommendations for Montevideo and Parque Rodó Neighborhood
Reprinted from: *Smart Cities* **2020**, *3*, 479–510, doi:10.3390/smartcities3020026 27

Irene Lebrusán and Jamal Toutouh
Using Smart City Tools to Evaluate the Effectiveness of a Low Emissions Zone in Spain: Madrid Central
Reprinted from: *Smart Cities* **2020**, *3*, 456–478, doi:10.3390/smartcities3020025 59

Víctor Manuel Padrón Nápoles, Diego Gachet Páez, José Luis Esteban Penelas, Olalla García Pérez, María José García Santacruz and Fernando Martín de Pablos
Smart Bus Stops as Interconnected Public Spaces for Increasing Social Inclusiveness and Quality of Life of Elder Users
Reprinted from: *Smart Cities* **2020**, *3*, 430–443, doi:10.3390/smartcities3020023 83

About the Editors

Luis Hernández-Callejo received a Ph.D. degree in the field of renewable energy and load forecasting from the University of Valladolid, Spain, in 2014. He is a Professor at the University of Valladolid, in the area of renewable energy and microgrids. He has supervised four Ph.D. students and has extensive research publication experience as an author, reviewer, and editor. He has participated in numerous research projects on energy, computing, and renewable energy. Currently, his research interests include renewable energy, microgrids, and the application of artificial intelligence in energy.

Sergio Nesmachnow is a Full Professor and Researcher at Universidad de la República, Uruguay. He received a Ph.D. in Computer Science from Universidad de la República (2010). His research interests include computational intelligence, metaheuristics, smart cities, high performance computing, and simulation. He is Editor-in-Chief of the International Journal of Metaheuristics and a Guest Editor for Springer, Elsevier, Oxford Journals, and MDPI. He has had more than 300 articles published in journals and conferences and has conducted more than 50 research projects.

Preface to "Mobility and IoT for the Smart Cities"

This is the printed edition of the Special Issue "Mobility and IoT for the Smart Cities", published in Smart Cities by MDPI. This book compiles relevant expanded versions of the best articles presented at the Second Ibero-American Congress of Smart Cities (ICSC-CITIES 2019, http://icsc-cities2019.com), held in Soria, Spain, in October 2019. The articles in this book span several of the main themes addressed in the conference.

The article "CITIES: Ibero-American Research Network for Sustainable, Efficient, and Integrated Smart Cities" describes a network for smart cities, uniting 56 institutions from 15 Ibero-American countries, supported by CYTED, the Ibero-American Program of Science and Technology for Development.

The next two articles present relevant studies of mobility and transportation systems, which are crucial for the development of sustainable smart cities. Transportation by trains is important for large cities, and in this regard, the work "Power Supply Solution for Ultrahigh Speed Hyperloop Trains" analyzes power supply alternatives for electric railways. On the other hand, the work "Analysis of Sustainable Public Transportation and Mobility Recommendations for Montevideo and Parque Rodó Neighborhood" provides an analysis and characterization of recent sustainable initiatives developed for public transportation in Montevideo, Uruguay.

Smart cities include many intelligent sensors and devices, and the Internet of Things (IoT) paradigm has been applied to improve the quality of life of citizens. The work entitled "Smart Bus Stops as Interconnected Public Spaces for Increasing Social Inclusiveness and Quality of Life of Elder Users" describes the development of an inclusive smart bus stop prototype and the use of information and communication infrastructures to build interconnected public spaces. In turn, sensors are increasingly being used for controlling and monitoring pollution and air quality in modern cities. The work entitled "Using Smart City Tools to Evaluate the Effectiveness of a Low Emissions Zone in Spain: Madrid Central" analyzes the situation of Madrid Central (Spain), a low emissions zone subject to controversy. Statistical and regression analyses are applied to evaluate the effectiveness of traffic reduction policies in lowering air pollution and outdoor noise.

All the topics covered in this book are of interest worldwide, but they are especially pertinent to Ibero-American countries, which are developing initiatives to contribute to the design of better, smarter cities, working together in the CITIES network.

Luis Hernández-Callejo, Sergio Nesmachnow
Editors

Editorial

CITIES: Ibero-American Research Network for Sustainable, Efficient, and Integrated Smart Cities

Sergio Nesmachnow [1,*] and Luis Hernández-Callejo [2,*]

1 Universidad de la República, Montevideo 11300, Uruguay
2 Departamento Ingeniería Agrícola y Forestal, Universidad de Valladolid, Campus Universitario Duques de Soria, 42004 Soria, Spain
* Correspondence: sergion@fing.edu.uy (S.N.); luis.hernandez.callejo@uva.es (L.H.-C.)

Received: 23 July 2020; Accepted: 29 July 2020; Published: 31 July 2020

Abstract: This article describes CITIES, the Ibero-American research network for integrated, sustainable, and efficient smart cities. General/specific goals of the network are commented, and participant members are introduced. The main activities developed within the network are described, including research, education, outreach, and dissemination. Finally, some key aspects of the current and future work are presented.

Keywords: smart cities; research network; energy efficiency; sustainability; mobility; IoT

1. Introduction

The paradigm of smart cities proposes taking advantage of information and communication technologies to improve the quality and efficiency of urban life [1]. The concept of smart city includes several forms of intelligent processes oriented to innovation and adaptation in order to provide better response to challenges of citizens. Furthermore, smart cities rely on the increasingly embedding of smart devices into traditional physical systems, using different sensor devices, such as mobile phones or Internet of Things (IoT) enabled domestic appliances. These devices generate large volumes of data that provide an unprecedented opportunity for extracting useful insights to improve decision-making processes to increase efficiency and improve the quality of life [2]. Several relevant application areas benefit from this new paradigm, including energy and sustainability, education, health, transportation, environment, business, and others.

In this context, this article describes CITIES, the Ibero-American research network for integrated, sustainable, and efficient smart cities, an effort developed in Iberoamerica to promote developing a methodology for strategic planning that helps bring the cities of the region to sustainability through cooperation and knowledge transfer between participating research groups and institutions. This'article presents a description of the main goals and activities of the CITIES network and the milestones achieved during the first two years of the initiative.

The article is organized as follows. Section 2 describes the CITIES network, its main goals and participants. Section 3 summarizes the main activities of the network. Finally, Section 4 elaborates on the impact of the developed activities and forthcoming events.

2. CITIES Network

This section describes the goals and participants of the CITIES network.

2.1. General and Specific Goals

The general objective of the CITIES network is to develop a strategic planning methodology that helps bring cities in the Ibero-American region towards sustainability through cooperation and knowledge transfer between the participating research groups and taking into account the own problems of the cities of the countries that are part of the proposal.

In addition to seek increasing efficiency and sustainability, CITIES recognizes that modern smart cities must be eminently integral. The concept of integrability is crucial, since cities become a multidisciplinary and totally heterogeneous scenario, both in terms of areas of interest and in the partial or global objectives of these areas. Therefore, the CITIES network intends that integrability is the cohesion element of the different fundamental characteristics of smart cities of the future. These characteristics are summarized in three main groups:

- Climate change and environment: Energy (energy efficiency and renewable energy); vulnerability to natural disasters, water, solid waste management, etc.
- Integral urban development: Territorial planning, mobility, security, connectivity, etc.
- Taxation and governance: Governance, cultural heritage, etc.

The specific objectives of the CITIES network include:

- Strengthen the technical and professional training of research centers and participating institutions and, therefore, contribute for the improvement of the quality of life and the environment of Ibero-American citizens.
- Foster new investment opportunities in targeted sectors that can lead to economic diversification and commercial opening of companies, through the identification of barriers and competitive advantages.
- Promote the environmental and economic benefits produced by the incorporation of CITIES among members of the public administration, group of architects, construction companies, financing institutions, companies, research centers, society in general, etc., by carrying out specific directed activities to these groups in a clear commitment to the sustainable development of cities.
- Search for strengths and weaknesses among the participating entities to consolidate a continuous and active exchange of scientific knowledge that allows the development of an action procedure in the cities of Ibero-America, as well as the formulation of future Research & Development projects that lead to consolidate a future line of work around CITIES.
- Promote meetings of researchers that integrate the participant groups at conferences and seminars, in order to identify potential projects for implementation in the following areas: Intelligent infrastructure, solutions for monitoring and preserving cultural and environmental heritage, applications oriented to efficient and sustainable mobility, promotion of renewable generation sources and energy efficiency in new smart cities, development of solutions for smart cities, and management based on IoT.
- Disseminate the results acquired during the project in usual venues (conferences, seminars, workshops, journal publications, etc.).
- Carrying out specific courses in the different areas of the project: Climate change and the environment, Energy efficiency and sustainability, Comprehensive urban development, Taxation and governance.
- Promote networking to address common problems to Ibero-American countries within the scope of the project. In the same way, the network will serve to identify new participating partners in the research and development, commercial, productive and institutional sectors, including the participation of end users (public administrations and companies).

2.2. Participant Institutions and Research Groups

A total number of 56 institutions from 15 countries participate in the network, from academic, public, and private sectors. Participating institutions include:

- Argentina (three institutions)

- o Universidad Nacional de La Plata
- o Universidad Gastón Dachary
- o Universidad Nacional del Litoral

- Brasil (three institutions)
    - o Universidade Federal da Paraíba
    - o Centro Federal de Educação Tecnológica Celso Suckow Da Fonseca
    - o ATIVA CITI
- Chile (two institutions)
    - o Universidad de Concepción
    - o Empresa Eléctrica ENEL Distribución Chile
- Colombia (eight institutions)
    - o Universidad Javeriana de Cali
    - o Universidad de Santiago de Cali
    - o Universidad Libre
    - o Universidad del Valle
    - o Universidad Nacional de Colombia
    - o Universidad de Antioquía
    - o CINTEL
    - o CIDET
- Costa Rica (one institution)
    - o Instituto Tecnológico de Costa Rica
- Cuba (one institution)
    - o Universidad Central "Marta Abreu" de Las Villas
- Ecuador (two institutions)
    - o Universidad de Cuenca
    - o Gobierno Municipal de Cuenca
- Spain (twenty-six institutions:
    - o Universidad de Valladolid
    - o Universidad Politécnica de Cataluña
    - o Universidad Carlos III de Madrid
    - o Universidad Pointificia Comillas
    - o Universidad de Salamanca
    - o Universidad de San Jorge
    - o Universidad de Zarragoza
    - o Universidad de las Islas Balearees
    - o Centro Universitario de la Defensa en la Escuela Naval Militar
    - o Diputación de Ávila
    - o AST Ingeniería
    - o Universidad del País Vasco
    - o Universidad Complutense de Madrid
    - o Centro Nacional de Energías Renovables

    - Centro de Desarrollo de Energías Renovables
    - Fundación CARTIF
    - CODESIAN
    - Ente Público Regional de Catilla y León
    - FORA Forest Technologies
    - INGARTEC
    - TECNALIA
    - TREELOGIC
    - VODAFONE Spain
    - ZIGOR
    - AYESA
    - CIEMAT
- México (three institutions)
    - Instituto Politécnico Nacional
    - Universidad Autónoma de Baja California
    - Universidad Nacional Autónoma de México
- Nicaragua (one institution)
    - Universidad Nacional de Ingeniería
- Perú (two institutions)
    - Universidad Nacional De Ingenieria
    - Universidad del Bio-Bio
- Portugal (one institution)
    - Instituto Politécnico de Bragança
- República Dominicana (one institution)
    - Instituto Tecnológico de Santo Domingo
- Uruguay (one institution)
    - Universidad de la República
- Venezuela (one institution)
    - Universidad de Zulia

Figure 1 presents the member institutions of the CITIES research network. Research groups in the project include more than 350 researchers and technicians from academic, public, and private sectors.

A committee of experts including 30 researchers from eleven countries, including expert from institutions outside the network, was appointed to help leading research and education activities within the network.

Figure 1. Member institutions of the CITIES research network.

3. Developed Activities

This section describes the main activities developed within the CITIES network.

3.1. Ibero-American Congress of Smart Cities

The Ibero-American Congress of Smart Cities (ICSC-CITIES) is the annual conference organized by the CITIES network, which gathers researchers, technicians, administration personnel and general audience to discuss project and initiatives related to energy efficiency, sustainability, mobility, governance, and other important subjects related to smart cities.

ICSC-CITIES is conceived as a discussion forum to disseminate both developed an ongoing research activities related to smart cities in the Ibero-American region. The main goal of the event is to create synergies among different research groups to foster the development of smart cities, and contribute to knowledge generation and integration in different scenarios, their possible development and the strategies to address them. The conference receives proposals of scientific articles based on the subject areas defined by the Committee of Experts: Energy Efficiency and Sustainability; Infrastructures, Energy and the Environment; Mobility and IoT; and Governance and Citizenship. The Technical Program Committee gathers more than 100 researchers from all over the world. Two editions of ICSC-CITIES have been organized (in 2018 and 2019), and a third edition will take place in 2020.

ICSC-CITIES 2018 was held in the Auditorium of Duques de Soria University Campus (Soria, Spain) in 6–9 September 2018. A total number of 101 articles were submitted to the conference, and a selection of the best 15 articles were published in Smart Cities, Communications in Computer and Information Science, volume 978, Springer [3]. Extended versions of the best articles were selected for consideration to be published in Revista Facultad de Ingeniería, Universidad de Antioquía, Colombia, indexed in Scopus, Emerging Sources Citation Index (Thomson Reuters-Web of Science), EBSCOhost, SCImago Journal Rank, LATINDEX, SciFinder.

ICSC-CITIES 2019 took place on Soria, on 7–9 October 2019. A total number of 100 articles were submitted to the conference, and a selection of the best 21 articles were published in Smart Cities, Communications in Computer and Information Science, volume 1152, Springer [4]. Extended versions of the best articles were selected for consideration to be published in Revista Facultad de Ingeniería,

Universidad de Antioquía, Colombia, and in the Special Issue "Mobility and IoT for the Smart Cities" on Smart Cities, MDPI [5].

ICSC-CITIES 2020 is scheduled to be held in Costa Rica, in 9–11 November 2020 (online venue, due to the COVID-19 pandemic). A selection of the best articles submitted to the conference will be published in Smart Cities, Communications in Computer and Information Science, Springer. Extended versions of the best articles will be selected for consideration to be published in Revista Facultad de Ingeniería, Universidad de Antioquía, Colombia and in the Special Issue "CITIES: Energetic Efficiency, Sustainability; Infrastructures, Energy and the Environment; Mobility and IoT; Governance and Citizenship", Applied Sciences, MDPI [6].

The ICSC-CITIES conference is open for researchers, students, academics, administration, and general public. No fees are charged for assistance, article presentation, or participation in tutorials. All accepted articles are publicly available in the proceedings [7,8]. All presentations, events, and discussion panels are recorded and publicly available at the conference website. The conference has been developed with the sponsorship of the Ibero-American Program of Science and Technology for Development (CYTED), Diputación de Soria (Spain), Ayuntamiento de Soria (Spain), Excelentísimo Ayuntamiento de Almazán (Spain), Universidad de Valladolid (Spain), Universidad de la República (Uruguay), Instituto Politécnico de Bragança (Portugal), Universidad Santiago de Cali (Colombia), Universidad Libre (Colombia), Springer Science+Business Media, and Grant Thornton.

3.2. *Development of Capacitation Programs and Courses*

A specific capacitation program is developed within the CITIES network, including presential and online training courses. In addition, workshops and seminars on several subjects of smart cities have been organized. These activities are conducted by specialists of the network and other relevant invited actors.

Several activities were developed in this concern, both in presential and on-line modalities, including:

- Presential seminar "Electric microgrids" at Universidad Santiago de Cali (Colombia), 2018.
- Presential seminar "Photovoltaic systems and their degradation" at Universidad Santiago de Cali (Colombia), 2018.
- Presential seminar at Instituto Politécnico Nacional (México), 2018.
- Presential seminar at Centro Federal de Educação Tecnológica Celso Suckow da Fonseca (Brazil), 2018.
- Presential seminar at Universidad de Cuenca (Ecuador), 2018.
- Tutorial course "Electric microgrids" at Universidad de la República, Uruguay, February 2019. Total participation: 200 assistants from academic, research, productive, university, and administration sectors.
- Presential course, Transient System Simulation Tool, Universidad de Valladolid (Spain), 2018 Armando Aguilar (UABC, México).
- Online course "Infrastructures, Energy and Environment", November 2019, 594 participants from academic, research, productive, and administration sectors.
- Presential seminar "Electric microgrids", Cali (Colombia), July/August 2019, 50 participants from academic, research, productive, university, and administration sectors.
- Presential course "Introduction to photovoltaic systems", Ciudad de México (México), June/July2019, 40 participants from academic, research, productive, and university sectors.
- Presential course, Transient System Simulation Tool, Universidad de Valladolid (Spain), November 2019, 10 participants from academic/university sectors.
- Presential seminary "Photovoltaic systems and its degradation", June/July2019, Cali (Colombia), 50 participants from academic, research, productive, private, and university sectors.

- Presential course "Mainteinance of photovoltaic systems", Ciudad de México (México), June/July2019, 40 participants from academic, research, productive, private, and university sectors.
- Participation in thematic seminar "Electric mobility", Madrid (Spain), June 2019, 20 participants from public and private sectors.
- Participation in seminar "Sustainable mobility", Madrid (Spain), December 2019, 20 participants from public and private sectors.

The courses and seminars have provided specific formation and training to more than 600 students and practitioners. A large proportion of the participants were post-gratuate student (M.Sc. or Ph.D. studies).

3.3. *Research and Development*

Research groups participating in CITIES network have developed valuable investigations in the research lines included in the initiative. More than 100 articles have been published in journals and conferences, and more than 25 corresponding to joint works between research groups participating in the network.

Some of the most relevant researches and published articles include the energy efficiency studies via computational intelligence for electricity demand forecasting in industrial and residential sectors [9]; IoT developments, such as the proposal of smart bus stops for increasing social inclusiveness and quality of life of elders [10]; electric mobility, such as a power supply solution for electric trains [11]; waste management in modern smart cities [12]; pollution and air quality, such as assessing the effectiveness of low emissions zones [13]; and sustainable mobility in public transportation [14].

3.4. *Dissemination and Outreach*

Several dissemination and outreach activities have been developed within the CITIES network. The most relevant participations in events include:

- Participation in ACELERATOR 2018 Business Architecture, Universidad Santiago de Cali (Colombia), November 2018.
- Participation and membership in Startup Cities Alliance (SCALE) since April 2019, by all members of the CITIES network. SCALE is an initiative of global cities represented by public builders, who work together on strengthening local ecosystems through sharing their networks and learnings.
- Participation in the round table about smart cities organized by ACCIONA and 20 minutos (Madrid, Spain) about Smart Cities, June 2019.
- Participation in the round table about "Cities and technology" organized by Universidad Social de Vallecas (Madrid, Spain) and Universidad Politécnica de Madrid (Spain). Invited talk "Integrated and sustainable smart cities".
- Participation in the presential workshop at Universidad de la República, Uruguay, "International Workshop on Transportation Planning and Smart Cities", February 2019, with the academic participation of 15 researchers and administration authorities from Uruguay, Argentina, Chile, México, and Spain.

3.5. *Search and Detection of Potential Smart Cities Projects*

This activity was oriented to identify current initiatives that can interact synergistically with the main research, development, and training activities under development in the CITIES network. After an exhaustive search 67 relevant projects in more than 30 countries were detected.

The analysis of related projects and possibly joint initiatives among institutions of the network has increased the proposal and development of research projects, agreements with authorities and institutions, and other joint ventures. In turn, the registration on M.Sc. and Ph.D. programs has also increased in the participant institutions.

3.6. Research Stays and Mobility

The CITIES network has promoted and funded research stays to develop joint works between research groups. A total number of 30 travel grants have been conceded, including travel for presential participation in the annual conference ICSC-CITIES. Most of the travels and research stays correspond to researchers from Latinamerican partners to develop activities in Spain. Participants from all countries of the network have developed mobility activities and research stays. In addition, 15 external participants from seven countries outside the network have developed mobility activities to participate in events organized by the network.

3.7. Meetings with Authorities and Other Institutions

A meeting was held with the Colombian Ministry of Education in 2018, and potential projects to be carried out with the Colombian Government have been identified.

The CITIES network was invited to participate in the Covenant of Mayors, an initiative of the European Union launched after the climate and energy package was adopted in 2008. The Covenant of Mayors co-operation movement involves local and regional authorities that voluntarily committed to increasing energy efficiency and the use of renewable energy sources on their territories.

4. Forthcoming Activities

The CITIES network is actively working in 2020 and has also specific activities planned for 2021. Regarding research, several on-going research and development projects are been developed and are planned to continue developing in the next year, including projects in the areas of energy efficiency, IoT and data analysis applied to public services, transportation design and planning, public services and governance, health, and others. A set of courses is also planned for 2020–2021, to expand the opportunities for formation of graduate and postgraduate students, and technicians.

Cooperation with authorities is also a crucial line of current and future work to guarantee the applicability of the proposed research and ideas by participating groups in the network. Contacts with stakeholders and companies in the sector are also been developed in order to foster new investment opportunities in targeted sectors. All developed networking activities also contributes to identify common problems and viable solutions to relevant issues in modern cities in Ibero-American countries.

The impact of the developed activities is shown in the high number of participants in courses, workshops, and conferences organized by CITIES. The number of joint research projects also demonstrates that the network is properly fulfilling the main objectives of promoting research and strengthen technical and professional training.

5. Conclusions

This article presented a description of the Ibero-American research network for integrated, sustainable, and efficient smart cities (CITIES), a research initiative to contribute with the development of smart cities in the Ibero-American region.

Both general and specific goals of the network were described. Institutions participating in the network were listed. The main activities developed within the network are described, including research, education, outreach, and dissemination. Finally, some key aspects of the current activities and main forthcoming activities were commented.

Author Contributions: Conceptualization, S.N. and L.H.-C.; writing—review and editing, S.N. and L.H.-C. All authors have read and agreed to the published version of the manuscript.

Funding: CITIES network is partly funded by CYTED, Ibero-American Program of Science and Technology for Development.

Acknowledgments: Authors acknowledge the contribution of all institutions and researches that are part of the CITIES network.

Conflicts of Interest: The authors declare no conflict of interest.

References

1. Deakin, M.; Waer, H. From intelligent to smart cities. *Intell. Build. Int.* **2011**, *3*, 140–152. [CrossRef]
2. Townsend, A. *Smart Cities: Big Data, Civic Hackers, and the Quest for a New Utopia*; W. W. Norton & Company: New York, NY, USA, 2013.
3. Nesmachnow, S.; Hernández Callejo, L. (Eds.) Revised selected papers. Communications in Computer and Information Science (CCIS). In *Smart Cities, Proceedings of the First Ibero-American Congress of Smart Cities, ICSC-CITIES 2018, Soria, Spain, 26–27 September 2018*; Springer: Berlin/Heidelberg, Germany, 2018; Volume 978.
4. Nesmachnow, S.; Hernández Callejo, L. (Eds.) Revised selected papers. Communications in Computer and Information Science (CCIS). In *Smart Cities: Proceedings of the Second Ibero-American Congress of Smart Cities, ICSC-CITIES 2018, Soria, Spain, 7–9 October 2019*; Springer: Berlin/Heidelberg, Germany, 2019; Volume 1152.
5. Nesmachnow, S.; Hernández Callejo, L. (Eds.) Special issue "mobility and IoT for the smart cities". *Smart Cities* **2020**, *3*, 2.
6. Hernández Callejo, L.; Alonso Gómez, V.; Nesmachnow, S.; Leite, V.; Prieto, J.; Ferreira, A. (Eds.) Special issue "CITIES: Energetic efficiency, sustainability; infrastructures, energy and the environment; mobility and IoT; governance and citizenship". *Appl. Sci.* **2020**.
7. Proceedings of First Ibero-American Congress of Smart Cities, ICSC-CITIES 2018. Available online: https://easychair.org/cfp/icsc-cities-2018 (accessed on 23 July 2020).
8. Proceedings of Second Ibero-American Congress of Smart Cities, ICSC-CITIES 2019. Available online: http://www.icsc-cities2019.com/ (accessed on 23 July 2020).
9. Porteiro, R.; Hernández Callejo, L.; Nesmachnow, S. *Electricity Demand Forecasting in Industrial and Residential Facilities Using Ensemble Machine Learning*; Revista Facultad de Ingeniería Universidad de Antioquia: Antioquía, Colombia, 2020.
10. Padrón Nápoles, V.; Gachet Páez, D.; Esteban Penelas, J.; García Pérez, O.; García Santacruz, M.; Martín de Pablos, F. Smart bus stops as interconnected public spaces for increasing social inclusiveness and quality of life of elder users. *Smart Cities* **2020**, *3*, 430–443. [CrossRef]
11. Lafoz, M.; Navarro, G.; Torres, J.; Santiago, Á.; Nájera, J.; Santos-Herran, M.; Blanco, M. Power supply solution for ultrahigh speed hyperloop trains. *Smart Cities* **2020**, *3*, 642–656. [CrossRef]
12. Rossit, D.; Toutouh, J.; Nesmachnow, S. Exact and heuristic approaches for multi-objective garbage accumulation points location in real scenarios. *Waste Manag.* **2020**, *105*, 467–481. [CrossRef] [PubMed]
13. Lebrusán, I.; Toutouh, J. Using smart city tools to evaluate the effectiveness of a low emissions zone in Spain: Madrid central. *Smart Cities* **2020**, *3*, 456–478. [CrossRef]
14. Hipogrosso, S.; Nesmachnow, S. Analysis of sustainable public transportation and mobility recommendations for montevideo and Parque Rodó neighborhood. *Smart Cities* **2020**, *3*, 479–510. [CrossRef]

Article

Power Supply Solution for Ultrahigh Speed Hyperloop Trains

Marcos Lafoz *, Gustavo Navarro, Jorge Torres, Álvaro Santiago, Jorge Nájera, Miguel Santos-Herran and Marcos Blanco

Unit of Electric Power Systems, Department of Technology, Centro de Investigaciones Energéticas, Medioambientales y Tecnológicas (CIEMAT), 28040 Madrid, Spain; gustavo.navarro@ciemat.es (G.N.); jorgejesus.torres@ciemat.es (J.T.); alvaro.santiago@ciemat.es (Á.S.); jorge.najera@ciemat.es (J.N.); miguel.santos@ciemat.es (M.S.-H.); marcos.blanco@ciemat.es (M.B.)

* Correspondence: marcos.lafoz@ciemat.es

Received: 30 April 2020; Accepted: 29 June 2020; Published: 9 July 2020

Abstract: The paper analyses the alternatives for the power supply of a Hyperloop type railway transport. The particular case of the technology of the Spanish company ZELEROS was studied. Based on previous technical specifications related to both the first prototype and a commercial system, different options were analyzed. We selected the use of a linear motor driven by a single power electronic converter, a distribution scheme comprising different sections along the acceleration area of the track, and an energy storage system based on supercapacitors for the energy supply. The power/energy ratio and the cycle capability are the reasons to become a feasible and competitive solution. A preliminary design methodology for the energy storage requirements is presented in the paper. Once the type of linear motor was selected, the power supply scheme was presented, based on a motor-side power electronic converter and a DC/DC converter which connects to the energy storage devices. An additional low power grid-tie converter for the recharge of the energy storage system was also used. Different track sections were defined, connected to the power electronic converter through corresponding switches, being supplied sequentially when the capsule presence is detected along the track. The particular characteristics of this application, with relatively short traction track area, as well as the high energy recuperation ratio due to the low losses, make more suitable the use of energy storage systems as the source of power supply than the direct connection to the grid.

Keywords: sustainable mobility; railway; Hyperloop; power supply; energy storage; power electronics; supercapacitors

---

1. Introduction: Historical References and Concept

In a global world, transport, both passengers and cargo, has become a strategic sector for the economy. That has implied an important increase in logistics, and transport has increased three times since 1950. Unfortunately, and as the European Environmental Agency has assessed, in 2016, the transport sector contributed 27% of total European greenhouse gas emissions [1]. Different European policies have established a reduction of them in two thirds by 2050 if compared with 1990 levels in order to meet the long-term 60% greenhouse gas emission reduction target as set out in the 2011 Transport White Paper.

Within this scenario of expansion of transport, congestion at the infrastructures, increase of CO_2 emissions, and dependence of fossil fuels is where the need for clean, sustainable, and efficient transport technologies is critical. As a result of these requirements, the concept of Hyperloop was launched in 2013 [2] as a new magnetically levitated ultrafast train, travelling along a tube under low pressure. Currently, there are several technological developments following the concept [3]. Some of them are in America [4], some others in Europe [5], and some in Asia [6].

The concept of Hyperloop could have some similarities with the very-high-speed trains developed in the 20th century, such as the magnetic levitation (MAGLEV) trains [7,8]. As a matter of fact, the first references about the Hyperloop concept are from the 17th century, when Denis Papin, a French physicist, envisaged the delivery of mail through pressurized air tubes. Later, in the 19th century, these ideas were materialized with the development of the London Pneumatic Dispatch Company (UK), presented in 1868 (Figure 1a). In 1867, in New York (USA), Alfred Ely Beach invented a similar device (see Figure 1b) which would become the first concept of suburban transport.

(a) (b)

Figure 1. (**a**) London Pneumatic Dispatch Company in 1868. (**b**) Pneumatic tube by Alfred Ely Beach, in New York (USA), 1867.

Nowadays, the scope of operation for this technology, both for freight railway and passenger traffic, is to connect distances in the range of 1000 km with the speed in the range of 700–1000 km/h. Several studies have been conducted recently in order to analyze the economic viability of this solution for both applications [9,10]. Within this speed range, the use of catenary and pantograph systems is not feasible because of the mechanical restrictions. Alternatively, the use of linear motors is preferred for traction.

Hyperloop technology is based on a railway line inside a pressurized tube where several moving parts, hereafter capsules or pods, are travelling with a frequency in the range of 2 and 6 min between capsules. Beyond these basics, different developers propose different technology details.

Figure 2 shows the technology by the Spanish company ZELEROS (Valencia –Spain-) [11], used in this paper as a case study to analyze the power supply in this application.

(a) (b)

Figure 2. Technology for a Hyperloop train from the Spanish company ZELEROS. (**a**) pod. (**b**) tube.

The solution proposed by ZELEROS comprises a rail line divided roughly into three areas, presented in Figure 3:

1. Firstly, an acceleration area, between 5 and 10 km, where the capsule is accelerated with the help of a linear motor installed along the rail. During this period, the capsule achieves a speed close to the maximum speed. Although different options are possible for the traction motor, it is preferred the one with less weight in the moving part and more simplicity in the coils supplied along the track line. Some details of the linear motor proposed are provided in Section 2.
2. Secondly, a cruise speed area, where the capsule maintains the maximum speed. The losses are compensated by means of an on-board propulsion system based on a reaction turbine. In the case of study, a compressor gets the air in front of the capsule, compresses it, and subsequently, it is expanded in the reaction turbine, in a similar way to an aircraft. This air is not only used for the propulsion of the capsule but also for the levitation in order to minimize the mechanical friction. The energy used for this process is covered by a set of batteries located on-board of the capsule. Obtaining the power supply for the on-board propulsion system is an additional technological challenge. However, they are out of the scope of this paper, and the details about it are classified. Nevertheless, it is worth mentioning that lithium-ion batteries have been proposed and analyzed as a solution due to the high energy density, low losses, and the appropriate power and energy ratio required for the track lengths under study. In this analysis, the pod is just a certain mass, which needs to be accelerated and decelerated. The weight of the batteries has been included in the total mass.
3. Finally, when the capsule arrives at the destination, it needs to be braked in a third area, named the deceleration area, using regenerative braking to recuperate part of the kinetic energy of the capsule. Additional braking systems are installed at the capsules in order to act in case of emergency. Due to the fact that most of the kinetic energy achieved on the travelling pods can be recuperated by regenerative braking, also considering that the losses are reduced because of the reduced pressure and levitation, the energy is likely to be reused for further acceleration of other pods travelling in the opposite direction.

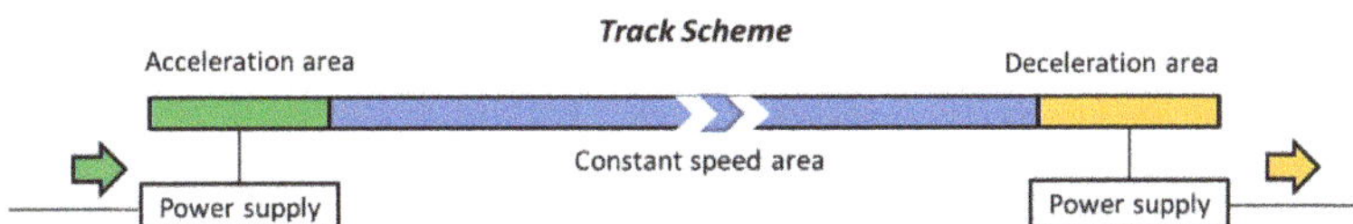

Figure 3. Track scheme with the different areas for acceleration, constant speed, and deceleration.

The paper aims at analyzing the different options for the power supply of this system, focusing on a specific technology for the linear motor. It is focused on the study of the power supply alternatives for the acceleration of the pods. Section 2 presents the system specifications according to which the power supply solution is defined. Section 3 presents the solution of linear motor used for this technology. Then, Section 4 analyses the different options for power supply in similar technologies. As a result, this section justifies the selection of an energy storage solution as a power supply for the system. Some calculations are additionally included to quantify the size of the energy storage requirements. Finally, Section 5 presents the practical implementation of a first reduced-scale prototype of the acceleration system, to be developed during 2020 and commissioned in 2021. Some particular scaling recommendations are included in order to get the maximum advantage of the reduced-scale prototype, considering the development of a commercial-scale version.

The main contributions of the paper are the proposal of an alternative solution for the power supply of Hyperloop railways, consisting of an energy storage system which supplies a linear motor drive during the acceleration of the pods. The impact of this solution could be high since it would allow the installation of Hyperloop stations mostly everywhere, without special requirements of the electric grid. The paper also presents a discussion of the different options and justification of the solution selected. Additionally, it presents a methodology to define and calculate the complete power supply system. It combines the design of the linear motor, the energy source and power electronics definition, and the distribution of the power supply along the acceleration area. Only a preliminary calculation of the energy storage system is provided, waiting to have more details about the energy losses related to the pod circulation along the low-pressure tube.

This paper is an extension of the work presented during the II Ibero-American Congress of Smart Cities (ICSC-CITIES 2019) celebrated in Soria (Spain) in October 2019, later selected to be published by Springer in the Proceedings of the Congress [12]. The CIEMAT Institute, the author of this work, is collaborating with the Spanish company ZELEROS in the development of the acceleration device and the power supply of the technological solution proposed.

2. Specification of the Power Supply for the Hyperloop System

The starting point of the study was to consider the technical specifications required by the system. They were provided by the company ZELEROS and compiled in Table 1.

Table 1. Technical specifications for the acceleration area.

Maximum Speed	Average Acceleration	Acceleration Time	Acceleration Length	Maximum Force	Mass
700 km/h (194 m/s)	2 m/s^2	97 s	10 km	100 kN	40,000 kg

One of the main challenges of the power supply of this transport system is that, and due to the ultrahigh speed, the moving part (pods) cannot be supplied by means of a conventional pantograph. This technology is only valid below 500 km/h because of mechanical limitations. Beyond that value, the solution of a linear motor deployed along the track is the most commonly used. An active part comprising electric coils is allocated on the track while the passive part of the electric motor is located on board. Additionally, such high speed would produce excessive friction between the moving part and the rail, and wheels cannot be used at that speed range. It is therefore mandatory to use some type of levitation in order to make the operation viable. Some examples are currently operating, such as the Transrapid and the Japanese MAGLEV JR-Maglev MLX, both magnetically levitated (MAGLEV).

All MAGLEV trains, as well as some of Hyperloop technological proposals, consider a linear motor, and therefore power supply, all along the track. That means a complicated set of high-voltage power lines and power electronic converters. As a consequence, there is an important constrain on the track length in order to keep the infrastructure cost within reliable limits. Some examples of power supply used for MAGLEV and Hyperloop solutions are presented in Figure 4.

On the contrary, the solution proposed by ZELEROS comprises: an acceleration area on the track, where a linear motor is installed and the pods are taken to the cruise speed; then a constant speed area of the track, where the speed is maintained by means of a propulsion system mounted on board and supplied by batteries; and finally a deceleration area of the track, where the pods are slowed down. This way, the longer the track, the more important the cost reduction of the power supply is in the total budget. Figure 4 presents a simple scheme of this structure.

Once the technical specifications for the dynamics of the system are selected, the first decision is to define the acceleration profile along the acceleration area. It is possible to choose a constant acceleration by applying a constant force, but it produces a very high maximum power level (around 15 MW in the study case); or, on the contrary, to select a more typical profile in industrial electric drives consisting of the first section of constant force and another section of constant power [13]. In this case, the maximum power is lower (10 MW for the case study) and therefore more convenient for the design of electric equipment, although it requires a higher force value in the first section. The second option is preferred since it gets better dynamics for the moving capsule and reduces the maximum power, as well as the current intensity, during the acceleration process. Force and power profiles will be later described in Section 5, where the design of the reduced scale is addressed.

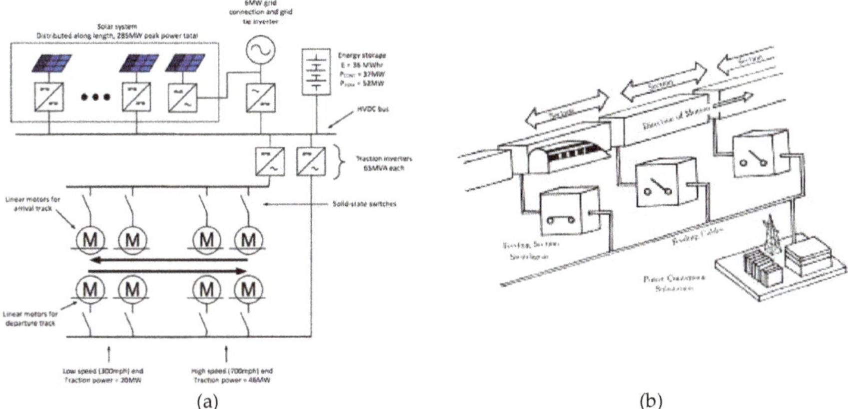

(a) (b)

Figure 4. Power supply schemes for: (**a**) Elon Musk's Hyperloop (**b**) MAGLEV JR-MLX.

The methodology to design the complete system is developed through the following steps:

- Based on the specifications and dynamic calculations, predesign of the linear motor and analysis of the electromechanical variables by means of finite elements (FEM).
- Based on the dynamics and basic data related to the track and conductors, selection of the number of power supply sections in order to fulfill the current and voltage requirements.
- Once the power supply sections are defined, the linear motor should be validated all along the acceleration area in order to check that the force is properly delivered during the whole range.

Unfortunately, the simple profile commonly used in electric drives is not optimal for this application. There are some areas along the acceleration section where the linear motor performance is insufficient and the back electromotive force is too high for the voltage available. As a consequence, the distribution of different types of coils along the acceleration area of the track is not uniform and leads to an asymmetrical deployment of the power supply and cabling. Authors believe that it would be very beneficial to modify the dynamic force and power profiles. An upgraded profile, where the power is composed of three sections, leads to a more homogeneous distribution of the coils along the track and to better use of the current levels. The implementation of this upgraded profile is presented in a case study included in Section 5 of the paper.

3. Selection and Design of the Linear Motor

The requirements to reduce at maximum the weight of the moving part lead to allocate the coils of the linear motor, named the active part, on the track [8]. Linear motion systems (LMSs) usually rely either on synchronous or asynchronous machines [13]. The former are far more widespread (Shanghai airport shuttle in China and JR in Japan), but the technology is expensive for long tracks; meanwhile, the latter are more economically competitive but have an excessive amount of losses on-board. As an alternative, the use of a switched reluctance machine is proposed. Despite having lower performance, it offers immense advantages as it saves costs and material while still allowing the system to transport high capacities at high speeds. Particular benefits of this linear motor technology: very light on-board part with low losses rate, robustness, high force density, and the possibility to operate in degraded mode when any of the phases fails [14]. An example of the electromagnetic analysis of the potential design of switched reluctance linear motor is presented in Figure 5. That solution corresponds to the particular case of the prototype defined in Section 5.

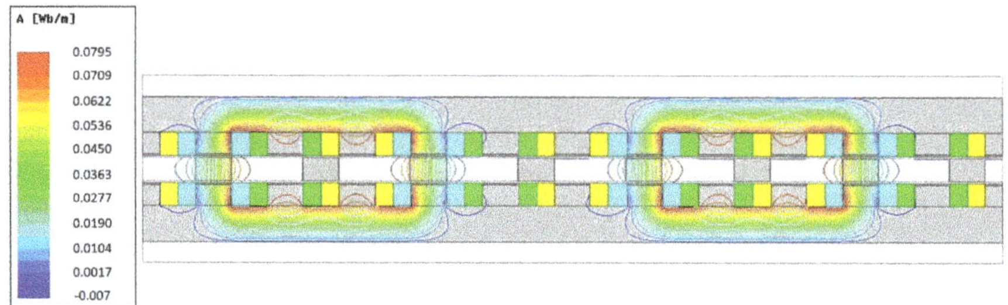

Figure 5. Electromagnetic analysis of a switched reluctance linear motor proposed for the propulsion.

A power electronic converter based on a three-phase full-bridge topology [15] was selected to drive this motor. The operation and the current waveforms of a similar switched reluctance machine (SRM) are presented in Figure 6, for the sake of example.

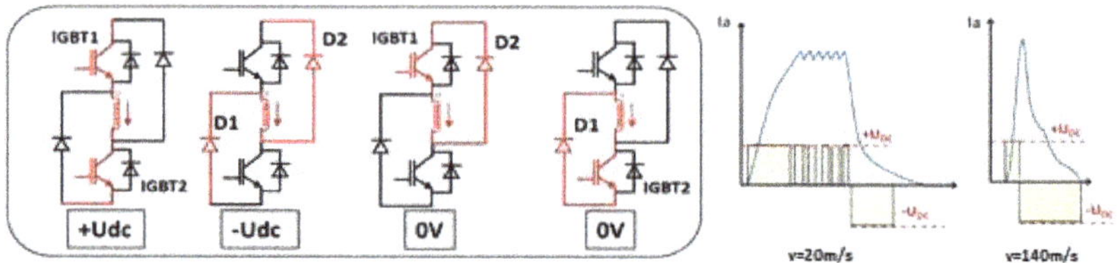

Figure 6. Switched reluctance machine (SRM) power converter topology and current waveforms for low speed and nominal speed.

4. Analysis of Options for the Power Supply of the Linear Motor

Once the linear motor and the power electronics are defined, the next step is to define how the system receives power and delivers it into the motor in order to accelerate it. Two basic options are proposed: direct power supply from the electric grid or power supply by means of an intermediate energy storage system.

4.1. Direct Connection to the Electric Grid

This scheme has the main advantage of allowing the use of high voltage, very convenient for the supply of the linear motor with many coils connected in a series in order to compensate for the voltage drop, allowing the use of lower current levels and therefore increasing the efficiency. However, the main inconvenience is a higher cost of a high-voltage infrastructure. Although no particular cost values could be gathered, it is well known that high-voltage equipment is in the order of 3 times more expensive (Red Eléctrica de España-REE). Moreover, the power level required, maximum of 10 MW, the relatively short length of the acceleration areas, around 10 km, and especially the short time of utilization of the power, in the order of 100 s, do not justify enough the investment in a dedicated infrastructure of high-voltage power supply.

As an alternative, it has been proposed to consider the use of energy storage systems, more flexible in utilization, without the requirement of a high power supply node.

4.2. Power Supply through an Intermedium Energy Storage System

The use of energy storage systems in the railway sector has some references, mostly in order to avoid the waste of energy during braking or to reduce the use of infrastructure. It is quite common to find light trains and trams with energy storage systems on board [16], more efficient than stationary solutions. However, the use of stationary energy storage devices is more convenient from the economic point of view, as well as to reduce the maximum power requirements to the grid [17].

4.3. Analysis of Energy and Power Requirements for the Energy Storage Device

According to the constant power profile from Figure 7b, the theoretical amount of energy required for the acceleration of the moving capsule (not considering the efficiency of the system or the additional power consumption of the auxiliary systems involved) with a maximum power of 10 MW can be estimated as the area under the red curve. A value of 217 kWh was obtained, as presented in (1).

$$\begin{aligned} E &= \tfrac{1}{2}\cdot(P_{max}\cdot t_1) + P_{max}\cdot\left(t_f - t_1\right) \\ &= \tfrac{1}{2}\cdot(10\ \mathrm{MW}\cdot 38\ \mathrm{s}) + 10\ \mathrm{MW}\cdot(97-38) = 780\ \mathrm{MJ} = 217\ \mathrm{kWh} \end{aligned} \tag{1}$$

Among the different energy storage technologies already existing in the market, three of them were selected according to their maturity level and the characteristics of high power, not very high energy and very fast response.

4.3.1. Batteries

The sector of the electric vehicle has triggered the development of batteries for this and other applications. Batteries are being applied for a wide variety of grid services. However, this technology presents two important disadvantages when considering an application like the one under analysis: the power/energy ratio is in the range of 1 MW/1 MWh, excessive energy for the required one (the case under analysis requires 10 MW and 0.217 MWh). Additionally, the number of charge-discharge cycles needed by this application is in the order of several hundred cycles per day, much higher than the cycles supported by batteries before suffering a loss of capacity (in the range of 5000–25,000 cycles). The use of batteries with such a high rate of cycles would lead to replacing the batteries every few months, resulting in a non-reliable solution.

4.3.2. Supercapacitors

Supercapacitors are a quite appropriate solution for the application in Hyperloop power supply because the power/energy ratio offered by commercial solutions is in the order of 1 MW/(5–10 kWh), very similar to the level required by this application. The main disadvantage is that commercial products present a voltage isolation limit of 1500 V for the series connection of cells. This restriction constrains the use for high power applications. Although using topology with a middle point grounded is possible to increase the voltage level to 3000 V, it is a technology challenge to go beyond this limit in order to allow the use in MW power levels without increasing the current values in consequence. The supercapacitor market trend goes towards getting cells with double the energy by 2020 and with higher isolation limits. If that is the case, supercapacitors would become very suitable for this application and economically competitive compared with grid connection or batteries. Some commercial products have been considered and most of them provide the same performance and characteristics [18]. The most restrictive parameter is the energy capacity, leaving the power capability underused.

4.3.3. Flywheels

Flywheels are also candidates because some characteristics (power/energy ratio and the number of cycles) are similar to supercapacitors. An additional advantage of this technology is that since power and energy are completely decoupled, commercial solutions are more flexible in terms of power and energy, and a more accurate design is likely to be selected to fit both parameters according to the application. However, the required high voltage level for the linear motor, in the order of 5000 V, has a negative influence in the feasibility of using flywheels, considering that the operation voltage of this technology is around 1000 V. It would require the use of a series connection, which would complicate the operation, or the development of high voltage adaptation through additional DC/DC converters, which would significantly increase the cost. As a result, this option was rejected.

5. Development of a 1/3 Reduced-Scale Prototype of Linear Motor and Power Supply

The first step before the development of a commercial line is the testing of a reduced-scale prototype to validate the technology. A 1/3-scale prototype of a linear motor and power supply system was selected in this case. The variables to be maintained from the initial design are: size and dimensions of the linear motor, force, current, and voltage. Only maximum speed is reduced, as well as the mass in order to achieve the maximum values with a much smaller track. The technical specifications for this system are presented in Table 2.

Table 2. Technical specifications for the acceleration area of the 1/3-scale prototype.

Maximum Speed	Average Acceleration	Acceleration Time	Acceleration Length	Maximum Force	Mass
500 km/h (139 m/s)	20 m/s^2	7 s	0.5 km	80 kN	2000 kg

Figure 7 presents the initial force (a) and power (b) profiles, according to the criteria defined in Section 1 of the paper. The requirements of the current and voltage are obtained from those profiles and therefore the design of both the linear motor and the power electronics can be accomplished.

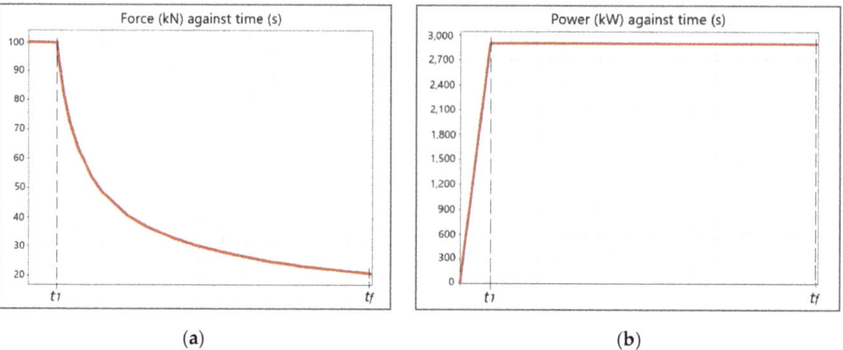

(**a**) (**b**)

Figure 7. (**a**) Force profile versus time. (**b**) Power profile versus time.

Starting from technical specifications and requirements for mass (m), maximum force (F_{max}), maximum power (P_{max}), maximum speed (v_f), and acceleration time(t_f), and having considered a constant power profile, it is necessary to establish the moment in which the force stops being constant. Considering in a preliminary calculation that the electric energy consumed by the linear electric motor is completely transformed into kinetic energy, Equations (2) and (3) provide the value of t_1, defined by (4), as the time when the system changes from a constant force profile into a constant power profile.

$$E = \int P{\cdot}dt \tag{2}$$

$$Ec = \frac{1}{2}{\cdot}m{\cdot}v^2 \tag{3}$$

$$t_1 = 2{\cdot}t_f - \frac{m{\cdot}v_f{}^2}{P_{max}} \tag{4}$$

For this particular case of reduced scale prototype, a three-segment power profile was defined, as introduced at the end of Section 2. A 10% higher value for the maximum power peak was defined in order to improve the performance of the system. Figure 8 presents the results, taking into account that the dynamics of the system are maintained in terms of journey time, maximum speed, and track length.

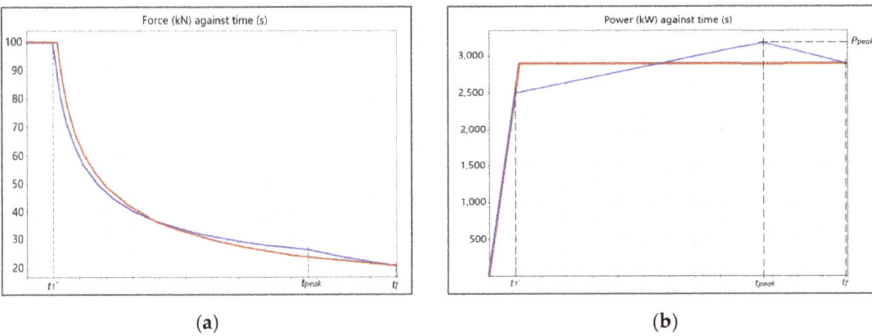

(a) (b)

Figure 8. (**a**) Force versus time profile and (**b**) power versus time for the 1/3 reduced-scale prototype.

Figure 9 presents the scheme of the complete system composed by the linear motor, the power electronic converter, the energy storage system, and the grid connection.

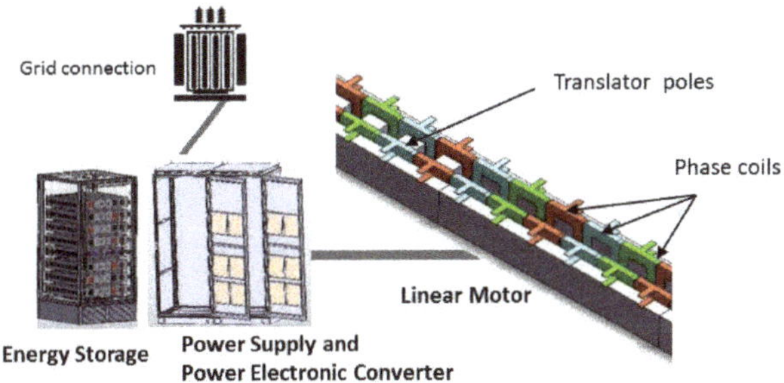

Figure 9. Scheme of the linear motor and the power supply system proposed.

5.1. Scheme of the Power Supply for the Reduced Scale System

The acceleration section needs a linear motor along 500 m in the reduced scale prototype. One of the linear motor design decisions is how to split the ampere-turns parameter into a number of coil turns and current level, taking into account the voltage drop and the power electronics design.

On the other hand, voltage requirements are also defined to compensate the electromagnetic force, the resistive voltage drop and the current transient over the inductance of the electric circuit, required to reach the current reference during the operation. As a result, the most suitable solution for the linear motor is to use different types of coils along the acceleration section. That is the best way to take advantage of the number of turns of the coil in order to adapt it to the back electromotive force (emf), according to the speed. The higher the speed the lower the number of turns of the coil. Five types of coils have been considered, with a number of turns varying from 12 to 4 when increasing the speed or the distance along the track.

The prototype under study comprises a power electronic converter with a voltage of 4000 V and a current of 3000 A. The limit on the voltage leads to defining different track sections, supplied independently and alternatively because the voltage is not able to cover the electromagnetic force, the resistance voltage drop, and the inductive effect of the whole linear motor. On the other hand, it is not necessary to supply all the coils along the acceleration area at the same time, just the section where the moving part is passing. The track sections group a number of linear motor coils, organized in what has been named unitary machines (UMs) in series. A UM is a set of coils corresponding to three phases and two pairs of poles. Every position on the track is equivalent to a force demanded, which is equivalent to a current value. Starting with the first track section, the number of UMs will be increased until a limit is reached in terms of voltage or current. Then, the first section is closed and a new track section is defined, following the same procedure. If a voltage limit is reached, a new track section with the same type of coil is considered. If the current limit is achieved, a new track section with a different type of coil is considered. A higher number of sections implies a higher number of switches to commutate the track sections and more cabling needed.

In order to reduce the number of sections, it would be convenient to increase the voltage level. Once the common voltage is reached, at the level of around 4000 V, new topologies such as multilevel power electronic converters [19] should be considered in order to manage voltage in the order of 6000 V with conventional semiconductors. This type of technology also allows to get additional values such as a reduction of the force ripple, reduction of the commutation losses, and more controllability in the current performance. Figure 10 represents the advantages of increasing the voltage from 4000 V to 6000 V by the available force along the track as a function of the number of unitary machines in a track section. The conclusion of these figures is that by increasing the voltage, the same extra force is available for a certain speed with a track section composed by roughly double the unitary machines, or double the number of coils. That means a lower number of sections and therefore less cabling and associated equipment for the track sections transition.

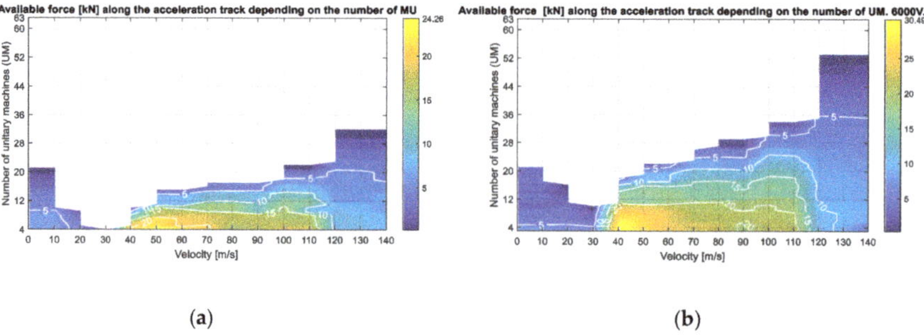

(a) (b)

Figure 10. Available force along the acceleration track as a function of the number of unitary machines (UM) included in a track section. (**a**) 4000 V power converter; (**b**) 6000 V power converter.

Force, and therefore current, is also related to the number of UM selected in each track section. Figure 11 presents an analysis of the total force, at different positions along the acceleration track, and depending on the number of UM selected. For a certain speed, the higher number of UM is included in the corresponding track section, the lower force is required to develop by each UM.

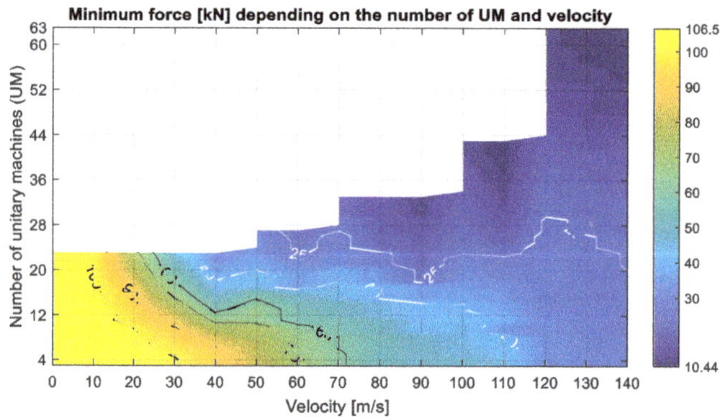

Figure 11. Minimum force required depending on the number of UM and velocity.

Another important issue related to the current is the selection of the current density at the linear motor coils (in order to determine the coils cable section) and the selection of the different track sections to be supplied at the same time by the power converter. Two current density values, 20 and 40 A/mm^2, were analyzed in order to consider two design criteria for the linear motor and presented in Table 3.

Table 3. The number of track sections with different current density.

Current Density	N = 12	N = 10	N = 8	N = 6	N = 4	Total Number
20 A/mm^2	2	1	2	6	17	28
40 A/mm^2	2	1	2	6	19	30

Table 4 complements the information of Table 3 with the length of the different track sections along the acceleration area for the case study using 20 A/mm^2.

Table 4. Length associated with the different track sections for the case of J = 20 A/mm^2.

Section	Coil Type	Start (m)	End (m)	Section	Coil Type	Start (m)	End (m)
1	N = 12	1	4	15	N = 4	192	216
2	N = 12	4	12	16	N = 4	216	238
3	N = 10	12	18	17	N = 4	238	261
4	N = 8	18	25	18	N = 4	261	284
5	N = 8	25	37	19	N = 4	284	306
6	N = 6	27	42	20	N = 4	306	329
7	N = 6	42	58	21	N = 4	329	350
8	N = 6	58	74	22	N = 4	350	372
9	N = 6	74	89	23	N = 4	372	393
10	N = 6	89	104	24	N = 4	393	414
11	N = 6	104	118	25	N = 4	414	436
12	N = 6	118	144	26	N = 4	436	457
13	N = 4	144	168	27	N = 4	457	478
14	N = 4	168	192	28	N = 4	478	500

Table 4 presents 5 types of coils with 4, 6, 8, 10, and 12 turns per coil, respectively. The coils with a higher number of turns are located at the first part of the acceleration area of the track (lower speed and therefore no need to reduce back emf), while the coils with a lower number of turns are located at the final part of it (higher speed and therefore need to reduce back emf). Skin effect might be also considered in this application since quite high frequencies are operating at the high-speed area. They were already studied in a previous analysis [12].

Position sensors will be deployed along the acceleration track in order to detect the presence of the pods. That will provide the closing of switches, connecting each track section to the power electronics converter. Only one track section is supplied by the power converter at the same time, only one power converter being necessary to supply the complete linear motor. That is an advantage compared to the power supply of previous similar concepts such as MAGLEV [20], where different power converters need to be installed along the track for the power supply, requiring high-voltage equipment. The switches selected must be semiconductors. They have to be fast, robust, and able to manage high voltage. However, they do not need to reach a high commutation frequency since they are only switched on once per trip. Therefore, thyristors are the best option.

5.2. Preliminary Design of an Energy Storage System for the Power Supply

Based on the system specifications, a preliminary design of an energy storage system was carried out. The methodology starts with a dimensioning based on the energy required, secondly based on the voltage, by connecting storage cells in series, and finally by current, validating that the current required does not overcome the maximum value of the cells.

5.2.1. Energy Dimensioning

The energy required during the acceleration process is based on the kinetic energy associated with the maximum speed of the pod. For this preliminary calculation, efficiency from electric to mechanic conversion was considered 100%, but it must be upgraded with the different efficiencies involved in the conversion process (aerodynamic losses, magnetic levitation losses, and Joule effect losses at the linear motor and the cabling along the acceleration area). The energy is given by (5).

$$E_{kinetic} = \frac{1}{2} \cdot mass \cdot v_{max}^2 = \frac{1}{2} \cdot 2000 \cdot 139^2 = 19.32 \text{ MJ} \tag{5}$$

Selecting one of the commercial modules with an energy of 80 Wh (64 V, 141 F) and taking into account a depth of discharge until half of the nominal voltage, which means 75% of the total energy, the number of modules required to fulfill the energy is 90, as it is obtained from (6).

$$\text{Number modules} = \frac{Energy_{required}(J)}{deep_{of_{discharge(p.u)}} \cdot Energy_{module(\text{Wh})} \cdot \frac{3.6 \text{ kJ}}{\text{Wh}}} = \frac{19.32 \text{ MJ}}{0.75 \cdot 80 \text{ Wh} \cdot 3.6 \text{ J/Wh}} \cong 90 \tag{6}$$

5.2.2. Voltage Dimensioning

The voltage depends on the maximum isolation voltage supported by the supercapacitor modules. Considering the same commercial 64 V modules, and taking into account a middle-point grounded topology (supports a maximum of 3000 V), the previously calculated 90 modules could be split into three branches of 30 series of connected modules each. This way, a maximum voltage of 1920 V is achieved. It remains the possibility to connect other 15 modules more in series per branch, increasing the energy of the complete system by 50% from the point of view of voltage limitation. The voltage of the storage system when reaching the discharge point is 690 V, the voltage used to calculate the maximum current.

5.2.3. Current Dimensioning

A higher current in the energy storage system is produced when the voltage is minimum. That implies 1429 A per supercapacitor branch, as obtained from (7) and once split into three parallel branches.

$$I_{supercaps\_P=cte} = \frac{P_{max}}{U_{supercap\_min}} = \frac{2.96 \text{ MW}}{690 \text{ V}} = 4289 \text{ A} \tag{7}$$

Taking into account that the maximum current of a supercapacitor module is 2000 A and that the average current during the operation is in the range of 1000 A, there is no limitation by the current. However, it is very important to analyze the thermal effects of the current since an increase of temperature over 60 °C could lead to an important loss of capacitance in the supercapacitors. A thermal model would be required to do an accurate study of the transient temperature evolution. However, considering that the operating time is only 7 s, and although this is a significant time from the point of view of the power electronics, it is not important for the thermal performance of the supercapacitors due to their relatively high thermal inertia. The already selected modules are therefore validated in terms of current.

It is important to add that the set of a series of connected supercapacitors will be connected to the linear motor power converter through a DC/DC converter. The recharge of the supercapacitors is done through a grid-tie converter (GTC), connected to the DC-link or directly to the supercapacitors, depending on the final voltage value, but of much less power than the DC/DC converter. A 50 kW GTC would be enough to replace the full energy of the energy storage system in less than 10 min.

6. Conclusions

The main contribution of this paper beyond the state of the art is to provide an alternative solution of power supply for Hyperloop railways, based on an energy storage system supplying a linear motor during the acceleration of the pods. This solution has the main advantage of being replicable in many locations independently of the power availability of the grid connection. That implies a big impact on the possibilities to deploy this transport. Additionally, a methodology to define and calculate the preliminary complete power supply system is also included in the paper.

The main results of this study are compiled as follows:

- The particular characteristics of a quite short acceleration section compared to the total track length for the application of a Hyperloop type railway, including a close deceleration section, implies that the use of energy storage is a viable solution for the power supply.
- It is not justified to install a high-voltage grid connection point for this particular case. On the contrary, the possibility to use an energy storage system provides replicability, independently of the grid connection, increasing the impact of the solution to almost any location. Moreover, the possibility to recuperate at the braking stage most of the kinetic energy as electric energy is an additional advantage in terms of energy consumption and sustainability.
- Once the power/energy ratio and the cycling requirements for this application are analyzed, it is obtained that the technology of supercapacitors results in the most convenient solution for this application. The potential to be associated easily in a series connection is an additional value in order to manage the use of higher voltages. A voltage of 4000 V is a recommended value to use for conventional power electronics. However, the need for higher isolation levels from the manufacturers, beyond the 1500 V, was addressed as a request in order to increase that voltage.
- The increase of operation voltage leads to better performance and a more convenient distribution of the power supply infrastructure with fewer track sections and therefore less cabling. It should be combined with the use of other power electronics topologies to increase the voltage, such as multilevel converters.
- The voltage limitation defined by the isolation voltage of the supercapacitor modules leads to high current levels. However, the use of different sections and the short periods of power supply contributes to a low increase in the temperature of the modules. The definition of the maximum current must be obtained from the study with a thermal model in order to determine the most appropriate current density to be used, avoiding a premature loss of capacity in the supercapacitors. A tradeoff between material cost and performance will be necessary.

The limitations found during this research were related to the lack of some experimental data. That is the reason why some of the calculations were just a first approach. The mechanical effects of the high speed can only be considered by testing in real high-speed conditions. That involves a quite large experimental set, not yet ready. As a consequence, some results were still approximated.

However, within the next steps, it is the development of a 1/3 reduced-scale prototype to test and validate the different parts of the technology. The first stage will be separate and conducted in the lab. The second stage will comprise the different parts together: linear motor, power electronic converters, and energy storage based on supercapacitors, and they will be tested in a 500 m testing track. The commissioning of this system is expected to be accomplished by 2021.

Author Contributions: Conceptualization and methodology, M.L.; design and calculation based on finite elements method, J.T.; modeling and control of the linear motor, M.B. and J.N.; calculation and analysis by means of simulation, G.N. and Á.S.; investigation, resources, and state of the art, M.S.-H. and J.N.; writing and reviewing, M.L. and G.N.; editing and visualization, J.T. and Á.S. All authors have read and agreed to the published version of the manuscript.

Funding: This research received no external funding.

Acknowledgments: The authors would like to acknowledge the support from the company ZELEROS by providing the specification and characteristics related to its product. It was essential for the development of this work.

Conflicts of Interest: The authors declare no conflict of interest.

Nomenclature

pod	Moving capsule along the pressured tube in a Hyperloop system.
MAGLEV	Magnetic levitation. Associated with the ultrahigh-speed trains.
JR-Maglev MLX	Japanese MAGLEV train.
SRM	Switched reluctance machine. A type of electric motor.
P_{max}	Maximum power demanded by the linear motor.
t_1	Time when the power profile passes from constant force to constant power.
t_f	Time of acceleration for the pod.
DC/DC	Direct current power converter. Connection of the energy storage system with the rest of the system.
F_{max}	Maximum force at the linear motor.
v_f	Maximum speed at the end of the acceleration area.
m	Total mass of the pod.
T_{peak}	Period when the peak power is achieved in a three-segment power profile.
emf	Electromagnetic force of an electric motor.
UM	Unitary machine. Group of motor coils corresponding to three phases and two pairs of poles.
Number modules	Number of supercapacitor modules.
$I_{supercaps\_P=cte}$	Current demanded to the supercapacitors at P=cte.
$U_{supercap\_min}$	Voltage at the supercapacitors at the maximum depth of discharge.
GTC	Grid-tie converter. Converter to connect the system to the grid for recharging the energy storage system.

References

1. European Environment Agency. Greenhouse Gas Emissions from Transport. Available online: https://www.eea.europa.eu/data-and-maps/indicators/transport-emissions-of-greenhouse-gases/transport-emissions-of-greenhouse-gases-11 (accessed on 12 January 2020).
2. Musk, E. Hyperloop Alpha. 2013. Available online: http://www.spacex.com/sites/spacex/files/hyperloopalpha-20130812.pdf (accessed on 1 November 2019).
3. Kale, S.R.; Laghane, Y.N.; Kharade, A.K.; Kadus, S.B. Hyperloop: Advance Mode of Transportation System and Optimize Solution on Traffic Congestion. *Int. J. Res. Appl. Sci. Eng. Technol. (IJRASET)* **2019**, *7*. ISSN 2321-9653. [CrossRef]
4. Hyperllop One. Available online: https://hyperloop-one.com/global-challenge (accessed on 18 January 2020).
5. Hardt. European Hyperloop Program. 2018. Available online: https://ec.europa.eu/eipp/desktop/en/projects/project-print-9401.html (accessed on 18 January 2020).
6. Hyperloop Transportation Technologies. Available online: https://www.hyperloop.global/about (accessed on 18 January 2020).
7. Ono, M.; Koga, S.; Ohtsuki, H. Japan's Superconducting MAGLEV TRAIN. *IEEE Instrum. Meas. Mag.* **2002**, *5*, 9–15. [CrossRef]
8. Ikeda, H.; Kaga, S.; Osada, Y.; Ito, K.; Mugiya, Y.; Tutumi, K. Development of power supply system for Yamanashi Maglev Test Line. In Proceedings of the Power Conversion Conference—PCC'97, Nagaoka, Japan, 3–6 August 1997; Volume 1, pp. 37–41.
9. European Union Agency for Railways: Hyperloop—An Innovation for Global Transportation? Available online: https://www.era.europa.eu/sites/default/files/library/docs/hyperloop_innovation_for_global_transportation_en_1.pdf (accessed on 18 January 2020).
10. Werner, M.; Eissing, K.; Langton, S. Shared Value Potential of Transporting Cargo via. Hyperloop. *Front. Built Environ.* **2016**, *2*, 17. [CrossRef]
11. ZELEROS Hyperloop Technology. Available online: https://hyperloopupv.com/projects/ (accessed on 18 January 2020).
12. Lafoz, M.; Navarro, G.; Blanco, M.; Torres, J. Energy Storage Systems for power supply of ultrahigh speed Hyperloop trains. In Proceedings of the II Ibero-American Congress of Smart Cities (ICSC-CITIES 2019), Soria, Spain, 7–9 October 2019; ISBN 978-958-5583-78-8.

13. Chevailler, S. Comparative Study & Selection Criteria of Linear Motors. Ph.D. Thesis, n°3569. Ecole Polytecnique Federal de Lausanne, Lausanne, Switzerland, 2006.
14. Torres, P.M.; Lafoz, M.; Blanco, M.; Navarro, G.; Torres, J.; García-Tabarés, L. Switched Reluctance Drives with Degraded Mode for Electric Vehicles. Modeling & Simulation for Electric Vehicle Applications. *INTECH* **2016**, 97–124, Chapter 5. [CrossRef]
15. Ellabban, O.; Abu-Rub, H. (Eds.) Switched reluctance motor converter topologies: A review. In Proceedings of the 2014 IEEE International Conference on Industrial Technology (ICIT), Busan, Korea, 1 February–26 March 2014.
16. Mwambeleko, J.J.; Kulworawanichpong, T. Battery electric multiple units to replace diesel commuter trains serving short and idle routes. *J. Energy Storage* **2017**, *11*, 7–15. [CrossRef]
17. Ceraolo, M.; Lutzemberger, G. Stationary and on-board storage systems to enhance energy and cost efficiency of tramways. *J. Power Sour.* **2014**, *264*, 128–139. [CrossRef]
18. Radcliffe, P.; Wallace, J.S.; Shu, L.H. Stationary applications of energy storage technologies for transit systems. In Proceedings of the EPEC 2010—IEEE Electrical Power and Energy Conference: "Sustainable Energy for an Intelligent Grid", Halifax, NS, Canada, 25–27 August 2010.
19. Fernão Pires, V.; Cordeiro, A.; Foito, D.; Pires, A.J.; Martins, J.; Chen, H. A Multilevel Fault-Tolerant Power Converter for a Switched Reluctance Machine Drive. *IEEE Access* **2020**, *8*, 21917–21931. [CrossRef]
20. Lee, K. Advances in the Application of Power Electronics to Railway Traction. In Proceedings of the 2015 6th International Conference on Power Electronics Systems and Applications (PESA), Hong Kong, China, 15–17 December 2015; pp. 1–4. [CrossRef]

 *smart cities*

Article

Analysis of Sustainable Public Transportation and Mobility Recommendations for Montevideo and Parque Rodó Neighborhood

Silvina Hipogrosso [1,*,†] and Sergio Nesmachnow [1,2,†]

1 Universidad de la República, Montevideo 11300, Uruguay
2 South Ural State University, 454080 Chelyabinsk, Russia; sergion@fing.edu.uy
* Correspondence: silvina.hipogrosso@fing.edu.uy
† These authors contributed equally to this work.

Received: 22 April 2020; Accepted: 21 May 2020 ; Published: 1 June 2020

Abstract: This article presents an analysis and characterization of recent sustainable initiatives developed for the public transportation of Montevideo, Uruguay. In addition, specific analysis and recommendations are proposed for the Parque Rodó neighborhood, based on a survey performed to people that commute to/from that area. The analysis considers the main concepts from related works, evaluating relevant quantitative (coverage, accessibility, affordability, etc.) and qualitative (public finance, integration, comfort and pleasure, etc.) indicators. Three sustainable public transportation initiatives are studied: electric bus, public bicycles, and electric scooters. Results of the analysis for each transportation mode suggest that the first initiatives focus on specific sectors of the population and should be improved in order to extend their accessibility and affordability. In turn, coverage must also be expanded. Regarding the analysis of the Parque Rodó neighborhood, results indicate that people are willing to perform the modal shift to more sustainable transportation modes, but several improvements are needed to improve the quality of service. All these aspects are considered in the proposed guidelines for a sustainable mobility plan in the area and also for suggestions and recommendations formulated to develop and improve sustainable mobility in Montevideo.

Keywords: sustainable mobility; public transportation; smart cities; mobility plan

---

1. Introduction

Mobility is a crucial component of modern society, where the participation of citizens in social, economic, and cultural activities requires traveling over both short and long distances [1]. Sometimes, traveling takes citizens a long period of time, regardless of the distance traveled, due to many reasons related to mobility situations. The ability of individuals to overcome limitations imposed by time, distance, and other mobility-related difficulties is critical to guarantee an active participation in city life [2].

Sustainable mobility is a subject that studies the development and use of transportation modes that are sustainable regarding several matters, mostly economic, environmental, and social [3]. Assessing sustainability and studying alternative transportation modes is very important considering that transportation largely contributes to environmental pollution with direct negative implications in health and quality of life of citizens. This is a relevant subject of study under the novel paradigm of smart cities [4].

Modern urban areas have been built around automobiles. This transportation mode has dominated the urban landscape, gaining the majority of the space on streets, limiting pedestrians zones, and reducing the space for other (sustainable) transportation modes. Private vehicles have revolutionized mobility, but they have also introduced several problems, including pollution, environmental degradation, congestion, accidents, a decline in public transportation, lack of accessibility, etc. However, nowadays many cities across the world are designing sustainable mobility plans to limit the use of private cars by improving public transportation, encouraging sustainable transportation modes, creating pedestrian zones, and changing the negative transformation caused by automobile dominance. While acknowledging the importance of car mobility in modern cities, this article focuses on analyzing sustainable transportation initiatives as an important mean of promoting the shift from cars to more sustainable transportation modes.

One of the most sustainable modes for mobility is provided by public transportation systems [5]. Public transportation allows moving a significantly larger number of people than private transportation, using a fewer number of vehicles. Furthermore, public transportation contributes to reduce greenhouse gas emissions, significantly improving the pollutant per passenger/km when compared with private vehicles. In the specific case of Montevideo, Uruguay, just a few initiatives have been recently proposed to promote sustainable private mobility (e.g., electric taxis and electric vans for last mile distribution of people and goods [6]). On the other hand, several recent initiatives have been proposed for sustainable public transportation, which are under development.

In this line of work, this article studies sustainable mobility initiatives recently developed in Montevideo, Uruguay: electric bus, public bicycles, and electric scooters. The main motivation of the study is to analyze and characterize the current reality regarding sustainable public transportation in Montevideo, in order to make mobility more sustainable. In turn, a specific analysis is performed on Parque Rodó neighborhood and Engineering Faculty, based on a survey performed to 617 citizens that travel from/to the studied area. Sustainable mobility alternatives are reviewed and categorized, and the main opinions about sustainable transportation modes are summarized and analyzed. Specific suggestions and recommendations are provided to develop and improve sustainable mobility in Montevideo and in Parque Rodó neighborhood.

This article extends our previous conference presentation "Sustainable mobility in the public transportation of Montevideo, Uruguay" [7] at II Ibero-American Congress on Smart Cities. New content and contributions in this article include (i) an expanded review of the related literature about sustainable mobility and proposals for Montevideo; (ii) the analysis of current sustainable initiatives in the public transportation system of Montevideo, regarding quantitative and qualitative indicators; (iii) suggestions and recommendations to develop and improve sustainable mobility in Montevideo; (iv) the analysis of the current situation regarding transportation and sustainable mobility in Parque Rodó neighborhood and Engineering Faculty; and (v) suggestions and recommendations proposed to develop and improve sustainable mobility in Parque Rodó neighborhood.

The article is structured as follows. Section 2 presents the main concepts related to sustainable mobility. A review of the main related work is presented in Section 3. The analysis of current sustainable mobility initiatives in Montevideo and specific recommendations are reported in Section 4. The analysis of the Parque Rodó neighborhood and Engineering Faculty, and the suggestions for developing and improving sustainable mobility in that area, are described in Section 5. Finally, Section 6 presents the conclusions and the main lines of future work.

2. Sustainable Mobility

Sustainability has been a major concern of modern society since the last decades of the twentieth century. Furthermore, sustainability has become a crucial aspect for nowadays communities, due to its direct implications on quality of human life and the growing awareness of the main issues and threats posed by environmental problems.

In 1987, the Brundtland Report for the World Commission on Environment and Development introduced the term sustainable development, to define "the development that meets the needs of the present without compromising the ability of future generations to meet their own needs" [8]. Sustainable development has become a paramount rule for modern sustainable mobility, i.e., the process to guarantee the movement of people with minimal environmental impact. Indeed, the World Business Council for Sustainable Development defined sustainable mobility as the ability of a society to fulfill requirements related to the movement of people without sacrificing fundamental human or ecological values [9]. Sustainable development is studied through three interrelated areas: the social, environmental, and economic dimensions. In several documents from the United Nations, sustainable development is said to be achieved when the goals of social equity, viable economic, and environmental friendliness are met in a coordinated manner [10]. Transportation and mobility are keys for sustainable development. Sustainable transportation can improve economic growth as well as improve accessibility (a very relevant social issue). Sustainable and safe transportation achieves a better integration of the economy while respecting the environment. As a consequence, sustainable mobility solutions must be designed considering these three areas, depicted in Figure 1, to contribute positively to their communities.

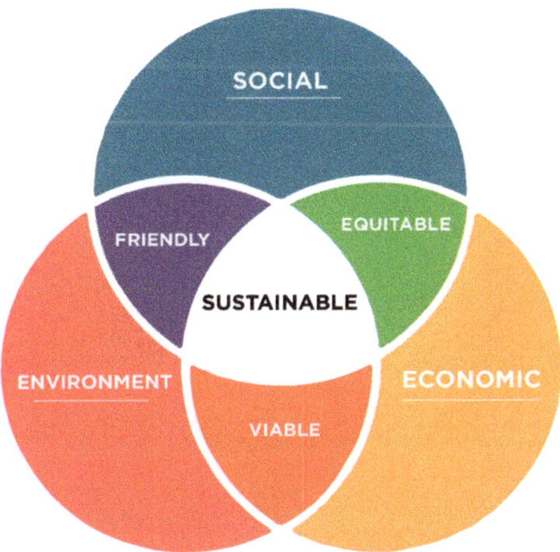

Figure 1. Main areas related to sustainable mobility.

The sustainable mobility paradigm integrates many relevant concepts [11]. Among them, a new approach has been proposed for designing and planning transportation systems, based on social processes, accessibility, reduction of motorized transportation, integration of people and traffic, and other factors to consider mobility as a valued activity, regarding environmental and social concerns [12].

Traditional research on sustainable mobility has focused on environmental impacts. However, several other important aspects have also been analyzed recently, including the relation with equity and the impact on economy, safety, health, and quality of life in general. In this regard, technology has been identified as one of the main tools that helps ensuring energy efficiency, using alternative and renewable energy sources, reducing contamination (e.g., pollutants emissions, noise, etc.), guaranteeing environmental friendliness, and providing tools for the analysis of reality and the development of new approaches based on data analysis. Furthermore, developing a rational plan for economic investments (e.g., infrastructure, transportation modes, etc.) is crucial to promote sustainable and solid growth in the long-term. Modern plans or courses of action are oriented to bring cities onto a more sustainable path with special emphasis on sustainable mobility, raising awareness, and involving citizens, with the main goal of fostering a behavioral change. The ultimate purpose is that citizens realize that transportation modes proposed by the sustainable mobility paradigm helps society, thus they opt for using more sustainable options by their own.

Many cities around the world aim at creating more compact, well-connected, and coordinated communities to reverse the development trend that focuses on private car and provides poor access to public transportation. Among several proposals for urban planning, Transit Oriented Development (TOD) stands as a model capable to contribute towards long-term environmental sustainability, in opposition to unrestricted growth usually happening in urban areas [13]. TOD promotes developing dense, compact neighborhoods that comprehend many uses (residential, commercial, cultural, institutional, etc.) physically connected by pedestrian and bicycle friendly transportation, also efficiently integrated with mass transit systems to connect with other zones within a city. The compact and integrated city design proposed by TOD also helps reducing carbon emissions, by providing people access to services at a walking distance. TOD also promotes socioeconomic benefits, expanding the connectivity of public transportation with housing and economic opportunities, which in turn enables developing more socially equitable cities by improving access to job opportunities and services.

Several indicators have been proposed and developed to study sustainable mobility in urban scenarios [14]. Among them, the most relevant are commuting travel time, coverage, access to mobility service, affordability, comfort and pleasure, intermodal connection, and integration. Indicators have been applied to analyze different transportation modes in many cities around the world. Some of the main related works on the topic are reviewed on next section.

3. Related Work

Modern cities are conceived to support different transportation modes, which are supposed to coexist, interoperate, and share the urban space. To guarantee a proper support for intermodality, transportation modes are supposed to be well integrated and connected, thus providing citizens with efficient and effective mobility [15]. However, the transportation planning problem is usually driven by traditional approaches that simplify, or even ignore, the complexity of handling several transportation modes coordinately. As a consequence, administrators often fail to provide holistic plans, accounting for all transportation modes operating in the city and including a comprehensive decision-making to consider indirect and interrelated impacts of the implemented solutions. Approaches that do not consider the city and transportation systems as a whole lead to isolated actions that usually result in poor and inefficient policies, which fails to solve the main problems related to mobility, among them, sustainability.

Litman and Burwell [16] established the relationships between sustainable transportation and mobility and recognized that in order to achieve sustainability, transportation must be conceived from a broad point of view to consider energy efficiency, health, economic and social welfare, and other relevant aspects related to sustainable development. Transportation impacts on sustainability were characterized into three broad areas (economic, social, and environmental) and it was stated that the correct approach for solving the underlying issues is to find strategies that help achieving all the main goals (in the long-term) by increasing transportation system efficiency. Several perspectives for addressing the sustainability problem in transportation were reviewed, and a list of common indicators for sustainable transportation was presented. Approaches for sustainability in transportation include improved travel choices, pricing and road design incentives, patterns for land utilization, and technical improvements to motorized vehicles, among others. A paradigm shift [17] was proposed for rethinking transportation, to consider different integrated solutions to achieve sustainable transportation systems.

Sustainable urban mobility planning begins with designing a strategic plan for the community. Banister [11] put special emphasis on stakeholder participation in the planning process, to involve them in the reasoning and implementation of specific initiatives for sustainable mobility. After questioning conventional principles for transportation planning, seven key elements for sustainable mobility were discussed and a more flexible paradigm to meet sustainable mobility purposes was proposed. Banister concluded that proactive involvement of relevant actors (including academia, policy-makers, and others) is crucial and more effective than traditional approaches for sustainable mobility. In this regard, specific changes must be made regarding land utilization, environment, public health, and ecology. Actors must accept their collective responsibility to achieve an effective sustainable mobility model to overcome car dependency. In particular, stakeholders must be engaged to support the application of measures towards promoting the main goals of sustainable mobility.

Methodological analysis and indicators have been studied as useful tools for the evaluation of sustainable mobility in cities [14,18,19]. Relevant studies are those related to understand the evolution of current transportation systems towards sustainable one, and those that evaluate the impact of selected solutions for specific case studies. In this context, indicators are used to simplify complex phenomena and to provide hints of different issues or problematic situations [20]. The combination of multiple indicators also allows capturing different dimensions and aspects of sustainable mobility.

The main concepts about indicators for sustainable transportation were presented by Gudmundsson et al. [14], focusing on the role and importance of quantitative and qualitative indicators for stakeholders (including decision-makers, planners, and operators). Frameworks for assessing sustainability metrics were reviewed, and a framework towards sustainable transportation was proposed. Two case studies were presented: European transportation and high-speed rail in England. The main conclusions from the case studies is that indicators and frameworks strongly depends on the context. Thus, understanding the context is key for succeeding in implementing effective sustainable mobility actions. Solid frameworks can stimulate politicians, decision-makers, and citizens to cooperate in the implementation of practical efforts regarding the studied topic.

Miller et al. [5] studied the role of public transportation regarding sustainability, reviewing related works that analyzed case studies of sustainable transportation. Frameworks, key challenges, and the benefits of public transportation were analyzed considering the three dimensions from sustainable development (environment, economy, and society). The reviewed articles were characterized regarding the main considerations for each of the studied dimensions. Finally, a set of recommendations were provided for developing and planning sustainable public transportation systems.

Rodrigues et al. [21] developed an index of sustainable urban mobility including several important features previously identified by Litman [17] for comprehensive and sustainable transport planning. The index is based on data obtained from planners and includes weights for different criteria, defined by experts. An application was presented for the city of São Carlos, Brazil, a medium-size city with 250,000 inhabitants. A basic analysis was proposed to demonstrate the viability and the efficacy of the proposed index. The main results of the study indicated that the proposed index was relatively easy to compute and flexible enough to be applied to characterize sustainable mobility.

Johnston [19] developed Production, Exchange, and Consumption (PECAS), a comprehensive method for modeling the impacts of transportation, to be included in an integrated urban model of California. PECAS was conceived as an spatial economic urban model, combining Walrasian concepts and random utility theory, using a network and a zone-based travel model to provide a theoretically valid measure of regional and statewide utilities. PECAS provides several data of economic utility like households by income, housing rents, housing affordability, etc. Several major transportation scenarios were studied, including metropolitan planning organizations of California, Sacramento, and San Diego. The metric evaluated included greenhouse gas emissions, air pollution, and economic welfare and equity. Results of the analysis were reported as relevant to state and regional transportation plans.

Baidan [22] studied the main problems of public transportation in Bucharest, Romania, which are similar to other capitals in Eastern Europe (e.g., the lack of policies to discourage car use, the lack of intermodal transportation options, etc.). Public transportation in Bucharest was analyzed through an accessibility indicator, considering the fares and the access of new residential areas to the transportation system. The main results of the study indicated that Bucharest has the cheapest public transportation fares among post-socialist European capitals, and that even though the metro does not cover many of the new residential areas, most of the citizens living there do have access to surface public transportation. A relevant suggestion was formulated, proposing the creation of intermodal hubs; networking the local/regional train systems with the metro; and providing other services as Park&Ride, bicycle parkings, rental services, etc., to promote sustainable transportation to citizens.

The successful case of sustainable mobility in Bogotá, Colombia, was analyzed by Lyons [23]. The study focused on specific actions oriented to address the protection of the environment and achieve both economic and social sustainability via a non automobile-centric approach. To satisfy the mobility needs, the city administration developed a mobility strategy based on three pillars: (i) discouraging the use of the car; (ii) the promotion of bicycle and walking as transportation modes; and (iii) the construction of an efficient public transportation network, anchored by a Bus Rapid Transit (BRT) system. Bogotá integrated transportation planning with social planning, by designing open areas and housing plans accessible to public transportation. According to Lyons, the results of the actions made in Bogotá span the three pillars of sustainability: economic, social, and environmental. Some of the outstanding results for the case study of Bogotá include BRT reduced travel times in 32%, 9% of former car drivers switched to BRT, traffic accidents reduced 93%, and air pollution decreased 40%. Overall, the study argues that the Bogotá case demonstrated that BRT is a viable and realistic solution as urban transportation plan, and also that increasing infrastructure for bicycle and walking is a viable strategy to promote sustainable transportation modes. The author concluded that the case study of Bogotá can be replicated in other developing countries, on the path towards sustainable transportation.

Moreover, in Latin-America, Rodrigues et al. [18] studied the provision of transportation services in Brazil following a specific methodology based on workshops with public managers and planners to characterize sustainable urban mobility. Multiple criteria analysis was applied to give support for decision-making, in order to identify proper goals, evaluate their relevance, and assess the impact of different solutions. Four stages were applied: (i) characterization of the problem; (ii) identification of relevant elements (including accessibility, congestion, infrastructure, social inclusion, pollution, non-motorized modes, and others); (iii) construction of a cognitive map, using operational and strategic concepts to reach the objectives; and (iv) identification of key viewpoints, using relevant concepts from the cognitive map, according to decision-makers. A series of workshops were performed in eleven Brazilian cities to gather information. Results were grouped by geographical regions, focusing on capturing different dimensions of sustainability in the context of each region, regarding three dimensions: social, economic, and environmental. The main identified problems were related to the relevance of urban public transportation, infrastructure, and environment. The importance of social, economic, and environmental issues reflected the development of each studied region. In conclusion, Rodrigues et al. stated the importance of the applied methodology to capture different views of sustainable mobility in Brazil and its application for creating public policies in that regard.

In Uruguay, project URU/17/G32 "Towards a sustainable and efficient urban mobility system in Uruguay" was launched in 2017, as a joint effort of government and transportation companies. The main goals of the project are defining regulations for low-carbon transportation systems, evaluating clean technologies in Montevideo, and promoting a cultural change towards sustainable transportation modes. Other recent initiatives for studying and developing sustainable transportation in Montevideo are project MOVES [6], which aims at promoting an effective transition towards inclusive, efficient, and low-carbon urban mobility in Uruguay, and project "Public transportation planning in smart cities" [24], funded by Fondo Conjunto de Cooperación Uruguay–México (2018–2019).

Regarding sustainable mobility plans for neighborhoods or specific zones of Montevideo, the case of Parque Rodó neighborhood presented in this article is the first study proposed so far. Current projects URU/17/G32 and MOVES have developed some initiatives regarding this subject, but focused on small and isolated cases, such as the Institutional Plan for Sustainable Mobility developed at the Uruguayan office of the United Nations Development Programme, with the goal of motivating people to adopt healthy mobility habits. This initiative is reported as "started", but no details about the methodology or results have been published yet. In any case, this is a very low-impact initiative, involving less than 20 persons that commute to the UNDP workplace [25]. Other minor initiatives related to sustainable mobility have been developed, but without proposing or implementing a sustainable mobility plan, such as replacing (non-sustainable) delivery vehicles for electric tricycles and electric pedal-assisted bicycles and incorporating electric utility vehicles for cargo companies.

Current project "Spatial, universal, and sustainable accessibility: characterizing the multimodal transport system of Montevideo, Uruguay" (code FSDA_1_2018_1_154502), funded by Fund for Research from Open Data, National Agency for Research and Innovation, Uruguay, proposes developing a characterization of urban accessibility, as an important tool to determine the quality and equity of transportation systems and the impact on daily activities of citizens. Several data sources are studied to characterize the accessibility of the transportation system of Montevideo. Three dimensions are considered in the analysis: territorial, universal, and sustainable accessibility. Territorial accessibility analyzes existing mobility alternatives to identify potential accessibility problems that prevent the participation of citizens in social and economic activities. The universal dimension studies accessibility problems of transportation systems for people with reduced mobility and the elderly, the alternatives of universally accessible transportation modes, and the identification of points of interest that are not universally accessible. Finally, regarding the sustainable dimension, sustainable mobility options offered by the city are studied, including non-polluting and low emission vehicles. A fourth dimension is proposed for the study, transversal to the previous three, to assess the impact of the socioeconomic condition on the three accessibility dimensions studied. It is expected that results from the project will be useful for the authorities, transportation companies, academia, and civil society in general, to identify situations of inequality in access to the different modes of transportation in the city.

The research reported in this article is a novel proposal for Uruguay, oriented to evaluating current sustainable mobility initiatives in Montevideo and Parque Rodó neighborhood. The research is part of current project "Spatial, universal, and sustainable accessibility: characterizing the multimodal transport system of Montevideo, Uruguay", developed within the sustainable mobility dimension.

4. Sustainable Public Transportation Initiatives in Montevideo

This section describes and analyzes sustainable mobility initiatives that are operating in Montevideo through public or private transportation companies.

4.1. The Public Transportation System of Montevideo, Uruguay

Public transportation in Montevideo is mainly comprised of buses. Taxis and other minor systems also operates in the city. Regarding the bus system, city authorities proposed the Metropolitan Transportation System (Sistema de Transporte Metropolitano (STM)), an urban mobility plan with the main goal of restructuring and modernizing public transportation in Montevideo [26]. Public transportation was integrated into a unified system comprised of 1528 buses operated by four private companies. The bus network consists of 145 bus lines, but considering the different variants of each lines, the number increases to 1383, a remarkably large number for a city like Montevideo.

The city center is a hub in the bus network, with most lines converging to that area. Additionally, the large length of certain bus lines with respect to the area of Montevideo is also noteworthy: the average median bus line length is 16.4 km, with the longest line spreading over 39.6 km, also a large number, as the area of Montevideo is 530 km^2.

In 2019, taxis were integrated in STM. Passengers can pay the rides on any transportation modes using the STM smart card, contactless top-up cards that are linked to the identity of the owner. STM can also used to pay for rides in the public bicycle system, as described in the next section.

4.2. Analysis of Sustainable Mobility Initiatives in Montevideo

Several sustainable mobility initiatives have been developed in the last years in Montevideo, including a prototype of an electric bus system (operating with just one line), a system for shared public bicycles, and a system for on-demand mobility using electric scooters. Other initiatives and measures to promote sustainable mobility and discourage the use of private motorized vehicles have been developed. The main details of each initiative are presented in the following subsections.

4.2.1. Electric Bus (Pilot Plan)

As in most countries in Latin America, Uruguay has recently prioritized moving towards cleaner energies in public transportation, in order to reduce its carbon footprint. For the last few years, transportation authorities in Uruguay have studied the potential benefits of including electric vehicles to the public transportation fleet in Montevideo. As a result, the Uruguayan government requested a loan from the Green Climate Fund, the entity that operates financial mechanisms to assist developing countries in adaptation and mitigation practices to counter climate change, in order to facilitate the modal shift from diesel to electric buses and allow Montevideo to replace 10% of the bus fleet [27].

The main public transportation company operating in Montevideo is CUTCSA, accounting for about two-thirds of the market share and also of the buses operating in the city [28]. CUTCSA has conducted tests of mobility using electric buses, with incentives and support from the Ministry of Energy and the city administration of Montevideo. Since 2017, a pilot plan is in course, using one electric bus that operates in different lines (the line changes weekly) to test the performance of this new transportation mode. The electric bus used in the pilot plan is a fully electric (no emissions) ByD vehicle, model K9A, with an autonomy of more than 250 km. It has an environmental friendly long-life iron phosphate battery of 324 kWh that admits more than 6000 charge cycles, and demands 3.5 h for total fast charge. The maximum speed of the bus is over 90 km/h and the average consumption is 100 kWh each 100 km. The pilot plan using electric bus has been considered an important advance for public transportation system of Montevideo. The new buses incorporates air conditioning to keep the environment ventilated, has decreased motor sounds and vibration, and offers universal accessibility, which is a great improvement in particular to those people with reduced mobility.

In 2019, 30 electric buses circulated on Montevideo with the purpose of gradually evaluating their operation and integration to the existing public transportation system. During 2020, CUTCSA plans incorporating 120 new electric buses to the transportation system of Montevideo [29].

4.2.2. Public Bicycles

The public bicycle system implemented by the city administration of Montevideo in 2015 is called Movete. It was conceived as part of the urban transportation system, in order to promote green mobility and a healthy way to know the city, move to workplaces and other relevant individual locations, and also to extend the accessibility of the public transportation system to specific final destinations.

As of December 2019, the public bicycle system consists of a fleet of 80 bicycles spread in a network of eight automated stations, distributed from the Old City to the Center neighborhoods. Bicycles can be rented at one station and returned in another station in the area covered by the system. A card of the integrated Metropolitan Transportation System (STM) is required to rent a bicycle. People that do not own a STM card, e.g., tourists or new users, can obtain it with no charge in the center office of Movete. The service has a time limitation for users, which cannot use the bicycles for more than four hours per day. This limitation is in line to promote more people have access to use the service.

The city administration is planning to expand the coverage (i.e., the area of operation) of the public bicycle service in 2020. The expansion is planned to include 60 stations and 600 bicycles, in order to increase accessibility and promote active mobility. New neighborhoods near the city center will be covered by the service, including Cordón, Parque Rodó, Parque Batlle, and Tres Cruces, accounting for a significant larger population then in the current implementation of Movete.

4.2.3. Electric Scooters

The electric scooter is a new mode of urban transportation that has gained popularity all over the world as an alternative to driving. Electric scooters provide an environmentally friendly alternative for short journeys that are either too far to walk, or too close to drive a car, to be a cost-effective option. Three companies of electric scooter (Grin, Lime, and Movo) operated in Montevideo since 2018, but two of them (Movo and Lime) stopped operating on December 2019.

Grin operates with an application where electric scooters and electric scooter stations are shown in a street map of Montevideo. The electric scooter stations are parking places where the company park scooters, with the permission of a close local business or an institution. Stations also provide a connection to the electric grid to charge scooters.

The service provides a practical and easy way to use electric scooters: by simply downloading a mobile application and setting up a payment method, users have access to a network of scooters that they can use at any time. Electric scooters have GPS blue tracking, so users are never too far from picking up a electric scooter and they can leave it anywhere within the area where the service operates.

To reduce logistic efforts of collection and distribution, Grin incorporated scooter stations after negotiations with drug stores, education centers, parking lots, and other commercial businesses around the city. Furthermore, the stations allow charging scooters. However, people still park scooters in any place. Up to December 2019, the company did not apply any penalty fee for not using the stations.

4.2.4. Other Initiatives

Other initiatives for sustainable mobility have been developed in Montevideo recently. There exist brand-new services that are not fully operative at the time of writing this article (February, 2020), or no data is available to perform a sound evaluation and analysis. Two of the main new initiatives are electric taxis and hybrid car sharing.

Electric taxis. The electric taxis initiative is in line with the main idea of government entities to promote using energy efficiency in public transportation, which have promoted the shift to more renewable energy sources, especially since 2015. As the change from gasoline to electric implies a higher initial investment, it is more profitable on taxis, which run many kilometers and the investment can be compensated with the lower operating cost. Currently, 54 taxis circulate on the streets of Montevideo and the main goal of the city administration is to increase the fleet of electric taxi to 300 vehicles (10% of the total fleet of taxis) by the end of 2020 [30].

Hybrid car sharing. The hybrid car sharing service provides citizens a mean to rent cars on-demand, for short periods of time. The system is accessed through a mobile application that allows users to choose from different locations to pick up and return the car, and the time they will use the service. The car sharing facilitator is a car brand (e.g., Toyota is one of the car brands providing this service in Montevideo). To promote the use of the hybrid car sharing service, the Uruguayan government has exonerated parking costs in the city for this transportation mode. Carsharing is a very new service in Montevideo (operating since late 2019), so there is not enough information or data to perform an in-depth analysis.

4.2.5. Promotion of Walking and Discourage the Use of Private Motorized Vehicles

Besides the aforementioned sustainable public transportation and sustainable mobility initiatives, other actions have been performed by the city administration to promote pedestrianism and reduce the number of private motorized vehicles in circulation in Montevideo.

Pedestrianism. Walking is the most sustainable transportation mode because is the only one that does not depend on any device or service. In Montevideo, multiple initiatives have promoted and stimulated pedestrianism; among the most relevant ones we can mention that walking lanes were incorporated in several parks, several streets in different neighborhoods (Old City, Reus, and even in low-income peripheral suburbs) were transformed for pedestrian-only use, and the constant reparation of sidewalks in a joint initiative (the Sidewalks Plan) with the participation of the municipality and residents. A relevant project to improve pedestrianism, which involved several infrastructure modifications, is the "Old city at human scale" project. Focused on the Old City neighborhood, the aim of the project is revitalizing the historic center of Montevideo, promoting sustainable mobility, universal accessibility, and improvement of public space [31]. Several important tasks were developed within the project, including repairing and widening all sidewalks, transforming them to single pavement; incorporating access ramps in every corner to improve accessibility; renovating the public lighting to improve safety; highlighting historical buildings and renewing urban equipment (street and square benches, litter bins, gardening, and signs, which were unified to give more information to the user); and building new rest areas, as suggested by neighbors. The project involved other activities related to the improvement of tourism service, environmental management, renovation of buildings that were badly damaged, and the incorporation of bicycle lanes all around the historic center. The project achieved positive results, thus improving accessibility, comfort on walking, safety, urban equipment on public spaces, and sustainability of the area.

Measure to discourage the use of private motorized vehicles. There is a rising concern about the impact of automobiles in the urban mobility area, which cause air pollution, human health effects, global climate change, congestion, and noise pollution. In order to achieve sustainable mobility, it is not enough to promote sustainable transportation modes; specific policies to limit the use of private motorized vehicles must be applied too. In this regard, the city of Montevideo has also proposed several measures to discourage the use of cars, thus contributing to sustainable mobility. Some policy measures applied include bus-only and preferential lanes for buses and taxis were deployed in many avenues, to promote the use of public transportation; tariff zones were applied for street parking in districts with high traffic and congestion (Old City, Downtown, and Cordón neighborhoods), the development of new sustainable transportation modes has been considered (e.g., an electric train connecting the city center with the Eastern part of the city) to reduce the dependence on automobile, the integrated public transportation system of Montevideo has recently incorporated taxis. All these measures are in line with the efforts to reduce car utilization. Furthermore, the administration has recognized the importance of promoting a cultural change, for citizens to do a responsible car utilization. Furthermore, project "Old city at human scale" also discourages the use of the automobile, taking away street space to give it to urban equipment and pedestrianization.

Other relevant measures, such as private traffic restrictions, which play a significant role in urban transportation regarding accessibility, air quality, and other factors that affect the quality of life of citizens [32,33], have not yet proposed nor implemented in Montevideo.

4.3. Indicators to Assess Sustainable Mobility in Montevideo

The proposed analysis considers a subset of sustainable mobility indicators proposed by the World Business Council for Sustainable Development [9]. The analysis applies a mixed approach, considering quantitative and qualitative indicators, which is the dominant methodology for sustainable mobility analysis, according to the review by Anagnostopoulou et al. [34]. On the one hand, quantitative indicators are those for which the available data for the case study allows computing a numerical value to assess a sustainable mobility criteria. On the other hand, qualitative indicators are metrics based on opinions, feelings, or points of view, rather than specific facts or numbers. Qualitative analysis are applied when the relevant pieces of information are not available for the studied initiatives. In particular, for the case of study of Parque Rodó neighborhood, a survey based on specific interviews to a sample of 617 persons moving to/from the studied area was performed. Resulting data were analyzed both using quantitative and qualitative indicators.

4.3.1. Quantitative Indicators

The quantitative indicator group includes coverage, access to mobility service, affordability, and commuting travel time. The corresponding definitions are presented next.

Coverage. The coverage is defined as the ratio of the area covered by each sustainable mobility service (*ci*) and the total urbanized area of the city (*ta*), according to Equation (1). The total urbanized area of Montevideo is considered to extend for 200 km^2. The scale for this indicator is straightforward, 0 correspond to 0% of coverage and 10 correspond to 100% of coverage.

$$coverage = \frac{ci}{ta} \tag{1}$$

Access to mobility service. Access to mobility service (*am*) is defined as the share of population with appropriate access to each service, according to Equation (2), where *nh* is the number of citizens living in the city, and *PR*(*i*) is the percentage of people living within 400 m from a public transportation stop or from a possible renting point of a shared mobility system.

$$am = \frac{\sum_i PR(i)}{nh} = 1 - \frac{\overline{PR}}{nh} \tag{2}$$

The methodology for calculation implies determining the percentage of people living within the service areas by using spatial data analysis. The service area is limited by a distance of 400 m of a sustainable mobility service, which is considered as the maximum distance that a person considers to walk to use a public transportation service [35]. The scale for the *am* indicator is 0 represents 0% of the population in the city and 10 represents 100% of the population.

Affordability of sustainable mobility transportation. Affordability (*af*) is defined as the expenditure on transportation made by persons as a percentage of their income. The calculation is based on the methodology by Carruthers et al. [36], considering the cost of performing 45 and 60 trips on each transportation mode and on existing socioeconomic data. The indicator is computed for two different relevant social groups, considering the minimum income and the middle income per capita, according to values reported for 2019 by National Institute of Statistics, Uruguay [37]. The calculation method is described by Equation (3), where *nt* is the number of trips, *p* is the cost of a single trip, and *is* is the income per capita. The scale for the *af* indicator is 0 indicates affordability index is over 35% and 10 indicates that is less than 3.5%

$$af = \frac{nt \times p}{is} \tag{3}$$

Commuting travel time. This indicator is defined as the average time spent by a person when travelling from origin to destination of a trip performed in the public transportation system. The methodology applied for calculation considers that (i) (for bus) the average commuting travel time includes the time for a person to walk to the bus stop and the time waiting for the bus to arrive; persons are supposed to walk from the centroid of the zone and the average walking speed is assumed to be 5 km/h; (ii) (for bicycles) the average speed is 13.5 km/h and the average walking time to a bicycle station is 4 min (walking up to 400 m); and (iii) (for scooters) the average speed is 12 km/h and the walking time a person takes to find a scooter is less than 3 min.

Commuting travel times are computed for two relevant distances: (i) a short travel of 3 km, a reasonable distance for travels to nearby locations such as offices, shopping, education, etc. It is also the average travel distance for electric scooters, considering an average speed of 12 km/h. (ii) A medium distance of 10 km, a reasonable average distance for travels to work, according to data from the urban mobility survey for Montevideo [38]. It is also the average travel distance on public transportation, considering an average bus speed of 13 km/h [39].

Two scales are considered for this indicator, for 3 and 10 km. Both consider as lower limit the time to travel the corresponding distance at the average human walking speed of 5 km/h, and as upper limit the time to travel the corresponding distance at the limit speed of bicycles and electric scooters (25 km/h). Thus, for the 3 km distance, 0 represent a trip duration of over 36 min and 10 represents a trip duration of 7 min, and for the 10 km distance, 0 represent a trip duration of over 2 h and 10 represents a trip duration of 24 min.

4.3.2. Qualitative Indicators

The qualitative indicator group includes net public finance, energy efficiency, intermodal connectivity, intermodal integration, and comfort and pleasure. The corresponding definitions are presented next.

Net public finance: Percentage of the cost of each mobility service that the government grants as subsidy to transportation companies.

Energy efficiency: Energy consumption in public transportation, usually evaluated in oil equivalent. The efficiency indicator considers the total energy demand from clean (i.e., renewable) and non-renewable sources.

Intermodal connectivity: Number of locations where users can change from one transportation mode to another.

Intermodal integration: Quality of the intermodal facilities between the different transport modes.

Comfort and pleasure: Satisfaction perceived by citizens about comfort and pleasure of moving in the city using different transportation modes. Comfort and pleasure indicator is analyzed through access to information, quality of the service, and security.

4.4. *Analysis and Results*

This subsection reports the results of the study to characterize the sustainable mobility initiatives. The study applies a urban data analysis approach, which has been also applied by our research group to study public transportation and other services in Montevideo [39,40]. The analysis accounts for relevant data about each initiative, obtained from open data sources (e.g., Open Data Catalog from the national government), data from previous studies (e.g., the urban mobility survey of Montevideo [38]), and also from personal interviews with both technicians of the local administration of Montevideo and managers of the companies that operate the studied initiatives (CUTCSA and Grin).

Quantitative Indicators

Coverage. The electric bus operated in several lines of CUTCSA company during 2017–2019. Table 1 summarizes the number of days of operation on the most relevant lines that operated the service. The percentage value for the number of days is also reported.

Table 1. Lines operated by the electric bus service in Montevideo (2017–2019).

Line	Days	Percentage
128	78	14.0%
142	16	2.9%
169	47	8.4%
180	303	54.4%
181/183 (circular line)	45	8.1%
187	20	3.6%
other lines	less than 6 days	less than 1%

According to the results in Table 1, the area considered to calculate the coverage of the electric bus service is the one corresponding to the buffer area defined by parallel segments located at 400 m of the most used lines routes: 128, 169, 180, and 181/183. The distance of 400 m is defined based on the recent mobility survey for Montevideo [38], which indicates that a person is willing to walk for up to about five minutes (corresponding to 400 m at a walking speed of 5 km/h) to access to a bus stop in order to use the public transportation service. In turn, for public bicycles, the coverage of the actual service and the projected coverage of the service are reported. The overall area for electric scooters is the one defined by the Grin service, which covers the area of service of the other two companies (Lime and Movo) that provided the service up to December 2019.

The area covered by each studied sustainable mobility initiative in Montevideo and the value of the *cov* indicator is reported in Table 2. Results were computed based on open data from each service.

Table 2. Coverage and the *cov* indicator for sustainable mobility initiatives.

Initiative	**Area**	**Coverage**	***Coverage* Indicator**
electric bus	51.4 km^2	25.7%	2.57
public bicycle	3.5 km^2	1.75%	0.175
public bicycle (projected)	13 km^2	6.5%	0.65
electric scooter (Grin)	23.5 km^2	11.75%	1.175
electric scooter (Lime)	15 km^2	7.5%	0.75
electric scooter (Movo)	7 km^2	3.5%	0.35
electric scooter (overall)	23.5 km^2	11.75%	1.175

The coverage maps for electric bus, public bicycles, and electric scooters services are presented in Figure 2. The analysis of the coverage indicator demonstrate that the area of service of each sustainable mobility initiatives is represents a small fraction of the total area of the city. The best coverage result was obtained for the electric bus service, which covers 25.7% of the city.

Coverage results are somehow expected, as the studied initiatives are new and public bicycles were introduced mainly for tourists. For electric scooters, coverage is also limited to zones with highest income (coastal area). Overall, the three studied modes provides a service that covers an area of 67.6 km^2, which represents 33.8% of the urbanized area of Montevideo, for a coverage index of 3.8. In conclusion, two-thirds of the the citizens who live in the urbanized area are not covered by these sustainable modes of transportation.

Access to mobility service. The population served by each service was computed by intersecting coverage areas with the population map and counting the total population in each zone. Figure 3 presents a superposition of the coverage area of sustainable mobility initiatives and the base map of population for Montevideo, grouped by census segments, which are used as the main administrative division for the Continuous Household Survey, from National Institute of Statistics, Uruguay [37]. The urban population of Montevideo is 1,305,082.

The electric bus service covers 429,269 citizens (32.9% of the population), accounting for the largest access index ($am = 3.29$). Public bicycles cover 86,917 citizens ($am = 0.67$) and the planned expansion is set to cover 193,368 citizens ($am = 1.48$). The electric scooters companies provides service to 285,445 citizens ($am = 2.19$). Overall, sustainable transportation modes cover 554,172 citizens (42.5% of the population, $am = 4.25$). As a consequence, the main conclusion from the analysis is that most of the urban population of Montevideo have no access to these sustainable modes of transportation.

Affordability of sustainable mobility. The affordability index was computed for the three studied transportation modes considering two types of trips: (i) short trips, with a length of 15 min, which is a reasonable traveling time for bicycles and and it is also the most frequent travel duration for scooters, according to the collected information, and (ii) long trips length of 45 min, which is the average time traveled in bus, according to the mobility survey [38].

Income per capita in Montevideo is USD 691 (middle income) and USD 423 (minimum income), as from data from August, 2019, and considering 1 USD = 37 Uruguayan pesos. On the one hand, electric bus applies a flat rate. The cost of a standard ticket, allowing one transfer trip in one hour, is 0.85 USD. On the other hand, electric scooters and public bicycles apply a time-based fare. The cost of using the public bicycles is 0 USD (free service) up to 30 min, and after that the rental cost is 0.74 USD for 30 min. For the electric scooter, the cost of a 15-min rent (the average time of utilization, as computed from the available data) is 2.1 USD and for one hour is 5.4 USD.

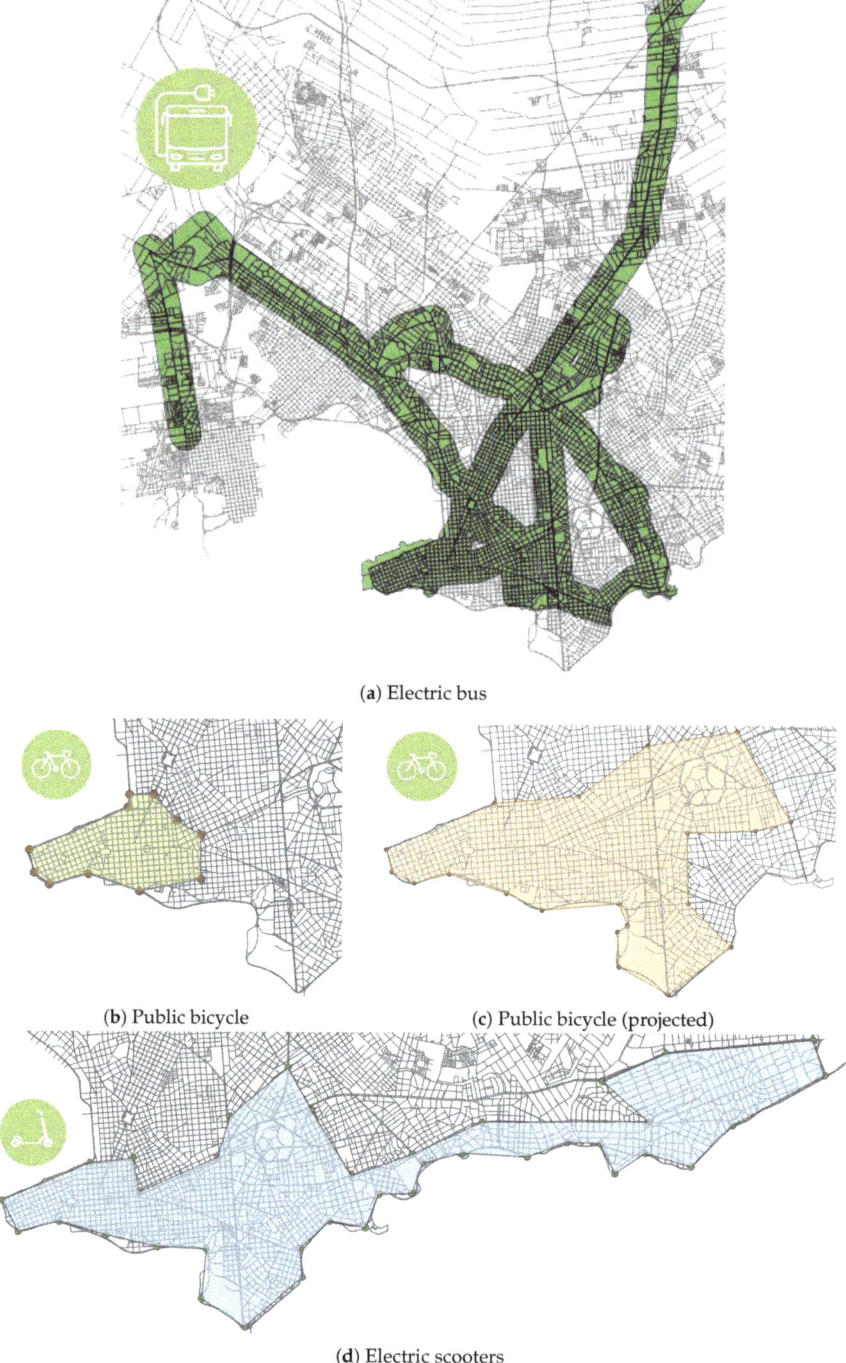

(**a**) Electric bus

(**b**) Public bicycle

(**c**) Public bicycle (projected)

(**d**) Electric scooters

Figure 2. Coverage of sustainable mobility initiatives in Montevideo.

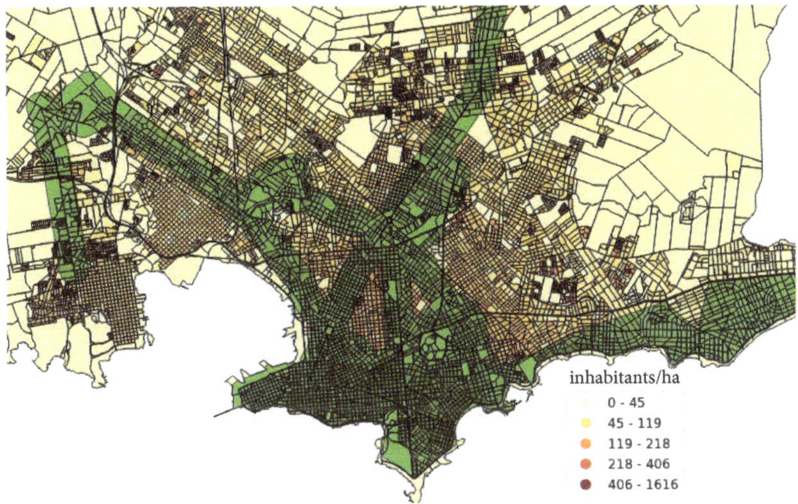

Figure 3. Coverage area of sustainable mobility initiatives and population of Montevideo (grouped by census segments).

Table 3 reports the affordability index of each sustainable transportation mode for middle and minimum income people.

Table 3. Affordability (*af*) indicator for minimum and middle income in Montevideo.

Trip Length: 15 min						
income	45 trips			60 trips		
	bus	bicycle	scooter	bus	bicycle	scooter
minimum	9.1% (8.2)	0 (10.0)	22.7% (3.9)	12.0% (7.3)	0 (10.0)	30.2% (1.5)
middle	5.5% (9.5)	0 (10.0)	13.9% (6.7)	7.3% (8.8)	0 (10.0)	18.5% (5.2)
Trip Length: 45 min						
income	45 trips			60 trips		
	bus	bicycle	scooter	bus	bicycle	scooter
minimum	9.1% (8.2)	4% (9.8)	57.2% (0.0)	12.0% (7.3)	5.4% (9.4)	76.2% (0.0)
middle	5.5% (9.5)	2.5% (10.0)	35.0% (0.0)	7.3% (8.8)	3.3% (10.0)	46.7% (0.0)

Results in Table 3 indicate that for 15 min trips, public bicycle has the maximum *af* value (10) for both income groups, as it is a free service up to 30 min. Affordability of bicycles does not reduce significantly when considering 45 min trips, due to the low fare of the service. Buses are cheaper than scooters for both short and long periods of time. Furthermore, the *af* indicator for buses is the same for both type of travels considered, while electric scooters downgrade to *af* = 0.0 for one hour trips. Overall, public bicycle is the most affordable transportation mode.

Commuting travel time. Table 4 reports the commuting travel times for three relevant distances for citizens' mobility in Montevideo: (i) 3 km, which is consider a short distance for those who primarily walk or ride a bicycle to work; (ii) 10 km, which is consider an average distance for bus commuters according to data from the urban mobility survey for Montevideo [38] and in line with similar studies for similar cities in the world [41,42]; and (iii) from end-to-end (*EtoE*) of the coverage areas for each mobility service. Speed and average times for bus were computed according to the methodology by Massobrio and Nesmachnow [40], using the Open Street Map service, estimations of average speed, and public applications available for the studied initiatives.

Table 4. Commuting travel times (min).

Bus			Bicycle			Electric Scooter		
3 km	10 km	EtoE (17.3 km)	3 km	10 km	EtoE (3.5 km)	3 km	10 km	EtoE (17.5 km)
17.8	49.3	116.0	13.3	44.4	15.6	17.0	52.0	89.5

Results in Table 4 indicate that bicycle is the fastest option for both short (3 km) and long (10 km) distances, followed by the bus, and in third place the electric scooter. Differences between bicycle and bus reduce for trips of 10 km. EtoE bus trips takes longer than traveling on scooter, and almost the same time for shorter distances.

4.5. Qualitative Indicators

Net public finance. The electric bus initiative has received benefits from three subsidies in order to reduce the ticket price: a subsidy from the city administration to implement reduced fees for students and retirees, a fuel subsidy from the Ministry of Transportation, and other contributions from the Ministry of Economy and Finance. Furthermore, in 2019, bus transportation companies were granted a total of 100,000 USD each to promote the substitution of 4% of diesel buses to electric.

The public bicycles service is completely financed by the city administration of Montevideo to promote active and sustainable mobility. Finally, electric scooters do not received any subsidy as they are run by private companies.

Energy efficiency. All the studied transportation modes use clean renewable energy. Public bicycle is the most efficient of the initiatives, as it does not requires energy of external sources. Electric buses provides a significant improvement over diesel vehicles regarding energy efficiency. They produce no CO_2 emissions and have an iron phosphate battery that consumes 100 KWh each 100 km, which is a good rate for public transportation. Regarding electric scooters, the energy of operation represents a very low percentage of the total emissions generated (e.g., 4.7% according to the study by Hollingsworth for the city of Raleigh, North Carolina [43]). However, several other concerns arise, such as the non-clean energy required for collecting and distributing scooters, and the short life cycle of batteries, which can have negative environmental impacts. Even though the company introduced scooter stations to avoid picking up scooters one by one, users continue leaving scooters anywhere (the company did not apply any penalty fee for not using the stations).

Intermodal connectivity. The studied sustainable mobility initiatives operate in a common area of 2.8 km^2 (considering the projected expansion for the public bicycles system, the area increases to 7.3 km^2). Within this common area, public bicycles offer full connectivity with buses and scooters, as stations are located less than 100 m of bus stops and scooters are available nearby. Electric scooters facilitate door-to-door mobility, allowing users to leave scooters in specific stations or even anywhere within the operation area, thus providing a valid alternative for intermodal connectivity. Buses also allows intermodal connectivity, but it is limited to a few bus stops that have bicycles or scooters stations nearby.

Intermodal integration. Even though the three transportation modes studied provide intermodal connectivity, the system as a whole lacks of intermodal integration. Each service focuses on their own operation, without facilitating integration with others: no information or route guidance is provided to users, terminal bus stations do not provide parking lots for public bicycles or scooters, etc. The only effective integration is regarding the payment method for buses and public bicycles, which can be paid using the same public transportation card (STM). All these facts are specific drawbacks for intermodal mobility. Overall, integration should be improved to provide efficient mobility.

Comfort and pleasure. Available information of public buses (e.g., via mobile applications) is recognized as one of the best features offered to citizens, according to the recent mobility survey for Montevideo [38]. On the other hand, trip comfort (43.9%) and bus stop comfort (46.4%) are the worst rated attributes of the bus system.

Users have presented claims about the poor service of Movete and bad conditions of bicycles [44]. Furthermore, Montevideo lacks of a proper infrastructure (e.g., exclusive bicycle lanes) for connecting stations of the system. Although the city administration planned to expand the network of bicycle lanes, even in the expanded configuration they will be not enough to properly satisfy the needs of an increasing number of users. In addition, it is difficult to complete even small infrastructure modifications, such as the case of the bicycle line in Parque Rodó neighborhood, which is commented in Section 5.

Finally, users perceive many benefits of electric scooters: they are easy to locate, ride effortlessly, dock-less, and can be parked anywhere. On the other hand, electric scooters are vulnerable to road risks, as they are driven on the same lane as automobiles, and are an uncomfortable transportation mode for bad weather conditions.

4.6. General Recommendations for Sustainable Mobility Initiatives in Montevideo

This subsection provides some recommendations and suggestions that can be implemented in the city of Montevideo to promote sustainable mobility. Recommendations and suggestions are based on the review, analysis, and main results of the study of the three initiatives for public sustainable mobility, reported in the previous subsection.

One of the main facts observed from the analysis is that the initiatives for sustainable mobility are not widespread through the city. Instead, they provide a limited coverage and poor access to citizens. In this regard, one of the main recommendation is related to expand the coverage area, by introducing more bicycle stations, operating new lines of the electric bus, covering different routes or extending the routes offered, and expand the areas available to operate electric scooters. To improve coverage, more vehicles must be introduced and an articulated network of exclusive lanes has to be designed and implemented, which will help to improve other indicators too.

Specific suggestions to increase accessibility are extending the bus and bicycles network, and also the electric scooters operation. The expansion requires a proper previous evaluation of the real demand for each transportation mode, via direct methods (surveys) and indirect methods (mobility data analysis approach). Another suggestion to increase accessibility is to perform a viability study of offering the studied mobility services to medium and low-income areas, thus increasing the social impact of the initiatives. The proposed suggestions are in line with the strategies for sustainable mobility by bus reviewed by Fernández and Fernández [45], and also with the development of similar initiatives in Latin America.

Concerning affordability, the study demonstrated that electric bus is expensive and electric scooters are prohibitive for low-income citizens. This is a critical issue, mostly considering the periodic fare increases for those services at least once a year. In this regard, a specific suggestion for mobility services is to provide ticket packages for frequent users, and offering a lower price for combinations with other services, to facilitate inter-modality. Public finance support can be reviewed to better contribute to affordability, mainly by redirecting the assistance to reduce operation and maintenance costs, to guarantee a lower price for each service.

Several other suggestions are related to improve travel time, in order to provide more useful and efficient sustainable transportation systems. In this regard, both city administration and transportation companies must focus on providing accurate information to citizens and guaranteeing a quick access to relevant information for travel planning. Electric bus should provide a higher frequency service, by redesigning or updating existing timetables, and a better effort must be done in order to provide good synchronization between different bus lines. For public bicycles and electric scooters, travel times are related to the availability of vehicles and also on the available interconnection network, so specific improvements on the fleets size and on infrastructure can contribute in this regard.

To take advantage of the modal shift from diesel to electric buses to improve energy efficiency, smart planning of battery charge is needed, by properly locating charge stations in strategic points of the operation area or planning the use of external batteries. Electric scooters also need to review their operation efforts for collecting and distributing vehicles, which currently demands non-clean energy. A specific suggestion to improve efficiency is installing secure parking stations to charge scooters batteries while parked.

A clear recommendation to enhance sustainable mobility is to promote intermodal connectivity between transportation modes. In this regard, services should work on providing real-time data information (e.g., vehicles available, location, bus stops information, timetabling, etc.) and on installing shared stations for at least two of the studied services. A specific suggestion is to integrate the ticketing system, allowing users to share modes within a ride, maybe linked with the aforementioned offers to improve affordability.

In terms of comfort and pleasure, companies can offer a better quality service by improving the comfort of the vehicles, and particularly adopting security measures to guarantee safe travels. Bicycles and electric scooters can incorporate helmets to their service and buses can include seat belts for passengers. Related to the overall quality of experience, companies and city administration can improve access to information providing users with mobile applications oriented to reduce walking time, waiting time, and the overall travel times.

Other mobility suggestions regarding relevant features such as age, gender, socioeconomic situation, etc. can be performed when proper data is available, in order to extend the overall analysis in the mobility survey [38]. We are working to get that information from the city administration (Intendencia de Montevideo) under current project "Spatial, universal, and sustainable accessibility: characterizing the multimodal transport system of Montevideo, Uruguay".

In general, the economic viability of the proposed suggestions is feasible within the current business models of the companies that operate each service. Furthermore, most of the suggestions are in line with current developments by national institutions (city administration, Ministry of Industry and Energy), which have committed funds for promoting and developing sustainable transportation and sustainable mobility in the city.

5. Practical Approach for Analysis and Implementation of a Sustainable Mobility Plan for Engineering Faculty and Parque Rodó Neighborhood, Montevideo

This section presents a specific case study that demonstrate the viability of analyzing and implementing a sustainable mobility plan for Engineering Faculty and Parque Rodó neighborhood, Montevideo.

5.1. Mobility Analysis and Survey

This subsection describes the studied area and the methodology for collecting and analyzing mobility data.

5.1.1. Engineering Faculty and Parque Rodó Neighborhood

Engineering Faculty (Facultad de Ingeniería) is the school in charge of engineering and other technology-related studies within Universidad de la República, Uruguay. In 2020, the Engineering Faculty has 10,350 students, 915 professors, and 195 administrative employees [46]. All these persons have specific mobility demands to access to the institution.

Engineering Faculty is located in Parque Rodó neighborhood (South of Montevideo). A map of the studied area is presented in Figure 4. The studied area covers 0.5 km^2 and includes three main avenues: Herrera y Reissig, where Engineering Faculty is located; Sarmiento; and Sosa. Nearby the Engineering Faculty is Aulario Massera, a large classroom building shared by Architecture, Economics, and Engineering faculties.

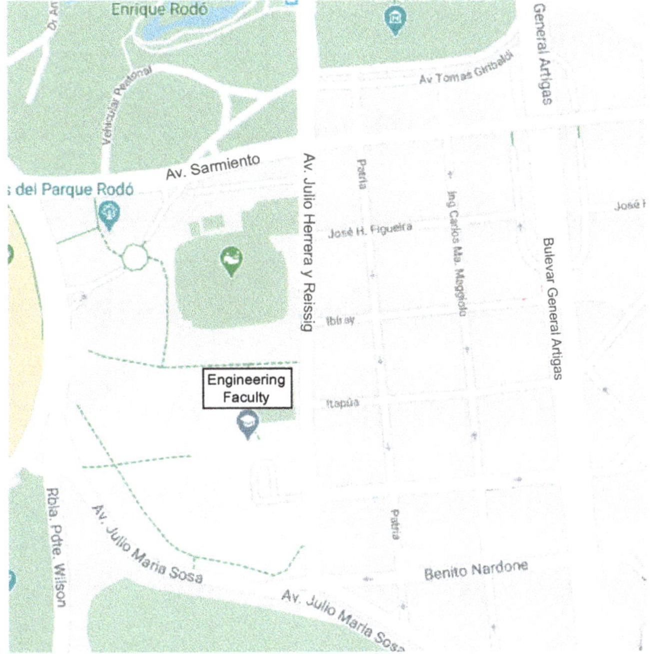

Figure 4. Area considered in the study: Engineering Faculty and Parque Rodó neighborhood.

Engineering Faculty has two parking lots with parking capacity for about 140 vehicles. The building also has bicycle parking (open from 7:00 to 23:00 from Monday to Saturdays) with security monitoring and a parking capacity of 330 bicycles. The bicycle parking has restrooms with showers and lockers to promote students using their own bicycles for traveling. This facility is under current norms for bicycles parking in public institutions, according to the administration of Montevideo.

Engineering Faculty has been promoting sustainable mobility initiatives. On June 2004, a group of professors founded "Unibici", a program to promote the use of bicycles between students. Moreover, Engineering faculty worked together with the city administration of Montevideo to create bike lanes in a circuit connecting faculties of Universidad de la República. However, the project has not been completed yet.

5.1.2. Motivation and Objectives of the Study

The main motivation of the study is to understand the mobility demands to Engineering Faculty and Parque Rodó neighborhood, and also from Engineering Faculty surroundings to other zones of the city. This is a relevant case study, which includes a variety of interesting features: Parque Rodó is a residential area, but also has a high education center (and others in the surrounding area), a shopping center nearby, several health centers in the zone, and other services. The objective of the study is to identify, analyze, and characterize the current situation regarding mobility and sustainable mobility in the studied zone, for different groups of people. This is a different study to the one performed to characterize sustainable transportation and sustainable mobility initiatives in Montevideo (reported in Section 4). In the case study in Parque Rodó, the study is based on data collected in situ and the opinions of interviewed people are taken in consideration.

The analysis of the current mobility situation provides quantitative and qualitative information for a systematic characterization of mobility demands in Engineering Faculty and Parque Rodó neighborhood. Moreover, the survey allows determining if the studied groups of people would be willing to change to more sustainable transportation modes and the specific issues that prevent them to make that change.

5.1.3. Methodology for Collecting Data

The methodology applied for gathering mobility information on the studied area consisted in collecting the information of the universe of study and performing a survey in situ.

Four relevant groups of people were identified: (i) students of Engineering Faculty and other faculties that shares Massera classroom building, (ii) professors and employees of Engineering Faculty, (iii) people who live in the neighborhood, and (iv) people who work on the neighborhood.

The total number of people involved in the analysis was 617 (79 living in the area + 538 commuting from other zones of the city). Thus, the study considered a sample size of 2.15% for the analysis of the mobility situation of Parque Rodó neighborhood. The estimated size of the relevant universe is 28,602 persons, including people that live in the studied zone; students, professors, and employees of Engineering Faculty; and persons that commute to the area from other zones of the city. The sample size considered in the survey is significantly larger than the one used is similar initiatives. For example, the mobility survey for Montevideo [38] studied 2230 homes, interviewing a total number of 5946 persons, which represent a sample size of 0.4% of the urban population of Montevideo.

By considering not only Engineering Faculty, but also the surrounding neighborhood, the survey intends to capture a more holistic view, taking into consideration the different groups of people that travels to/from the studied area.

A survey was formulated to know the mobility characteristics of the studied groups of people. The survey included the following questions.

1. Do you study or work at Engineering Faculty?
2. Do you travel often to this area?
3. What is the origin and destination of your trip?
4. What transportation mode(s) do you use for commuting to study/work in the neighborhood or from this neighborhood to other zones of Montevideo?
5. If you use more than one transportation mode, specify the percentage of utilization.
6. How often do you make these travels weekly?
7. Which aspects are the most relevant for you while commuting?
8. Would you be willing to switch to a more sustainable transportation mode?
9. To what transportation mode would you be willing to change?
10. What do you think it prevents you to change to a more sustainable mobility ?

The survey was performed face-to-face to people circulating in the studied area. Interviews were performed in different locations, including the front door of Engineering Faculty, five bus stops located less than 300 m of the faculty, a bakery located 100 m from the faculty, the front door of Franzini football stadium, and also in random locations at streets in the zone: Julio Herrera y Reissig, Itapua, Ibiray, Patria, José Figueira, Eduardo Garcia de Zuñiga, Benito Nardone, Julio Maria Sosa, Carlos María Maggiolo, Sarmiento, Senda Nelson Landoni, and Bulevar Artigas. People were not interviewed at home, because the main interest was in specific mobility demands (e.g., people attending to Engineering Faculty, moving from/to work, or moving to shops in the area).

The questionnaires were performed during 15 November–15 December 2019, from Monday to Friday, from 8:00 a.m. to 7:00 p.m. Weekend trips were not considered in the analysis because they are significantly lower than working days trips. Engineering Faculty offers just a few classes on weekends (just on Satuday morning, for some sporadic activities) and commercial activity in the studied zone is also reduced on weekends. People who commute in sustainable transportation modes were not asked if they would be willing to change towards a more sustainable transportation, as they already do it. The study also gathered information of bus lines that operates in the zone and identified the bus stops near the faculty. Scooter stations and bicycle lanes were also identified.

5.1.4. Methodology for Data Analysis

The study applies a urban data analysis approach, accounting for relevant data from the survey and also information from public sources.

Regarding the methodology applied for data analysis, the study analyzes global characteristics of mobility demand in the area. Some indicators used for the global case of Montevideo are studied, e.g., coverage and commuting travel time, as defined in Section 4.3. In addition, other relevant aspects related to the sustainable mobility characterization are analyzed, such as travel distance and modal-choice preferences for trips. Travel distance is defined as the distance that a person travels from any point of the city to the centroid of the Parque Rodó neighborhood. All distances are computed using the Google Maps service. Modal-choice preference of commuters is defined as the decisions taken by individuals to chose one transportation mode instead of another. The reason for the choice is linked to several factors, including affordability, travel time, comfort, accessibility, and sustainability.

Furthermore, the study analyzes the quality of service of existing mobility options through mobility preferences while commuting, such as cost, comfort, speed, security, sustainability, and other valuable interests for citizens. The studied mobility preferences may not correspond to the transportation mode that people use today, but to modes that they are willing to use if those preferences and related issues improve.

Some indicators analyzed in the case study of Montevideo are not taken into account in the study of Parque Rodó and Engineering Faculty. For example, affordability or access to mobility service indicators are not computed, mainly because of two reasons: (i) from the point of view of the price of mobility services, prices are the same for all zones in Montevideo, thus the main results reported in Section 4 also holds for Parque Rodó and Engineering Faculty neighborhood, and (ii) most of the studied universe consists of middle/high income people, which normally can afford all transportation modes (this fact is confirmed by the low number of trips from/to those zones of the city with the lowest income per capita, which is below 8%).

5.2. Analysis of Results

This subsection reports and discusses the most relevant results of the study. The most relevant results of the survey are presented on graphics, tables, and maps that allows characterizing distances, transportation modes, and other relevant features related to sustainable mobility in the studied zone.

Coverage. The studied area is fully covered by all the studied transportation modes (bus, bicycles, and electric scooters). Seven bus lines operates in the neighborhood, directly connecting people with many zones in the city. Furthermore, all locations in the city can be accessed via transfer trips. Although bus-only lanes were defined in main road and avenues of Montevideo, they are not defined in the studied area, so buses share the road with private transportation.

However, just 47 trips of the electric bus (3.6% of the total trips performed in 2016–2019) operated in lines that serve Parque Rodó neighborhood. Regarding bicycles, the current public system does not cover the studied area, but it is projected to be covered in the expansion, as reported in Section 4. Engineering Faculty provides the bicycle parking and other services for students, professors, and workers that use this transportation mode. Scooters operates all through the zone, having five stops near Engineering Faculty. Figure 5 present a coverage map of the studied zone, highlighting bus line routes and stops, scooter stops, bicycle parking areas, and bicycle lanes. The bicycle parking of Engineering Faculty is distinguished as it provides covered parking, security, and showers.

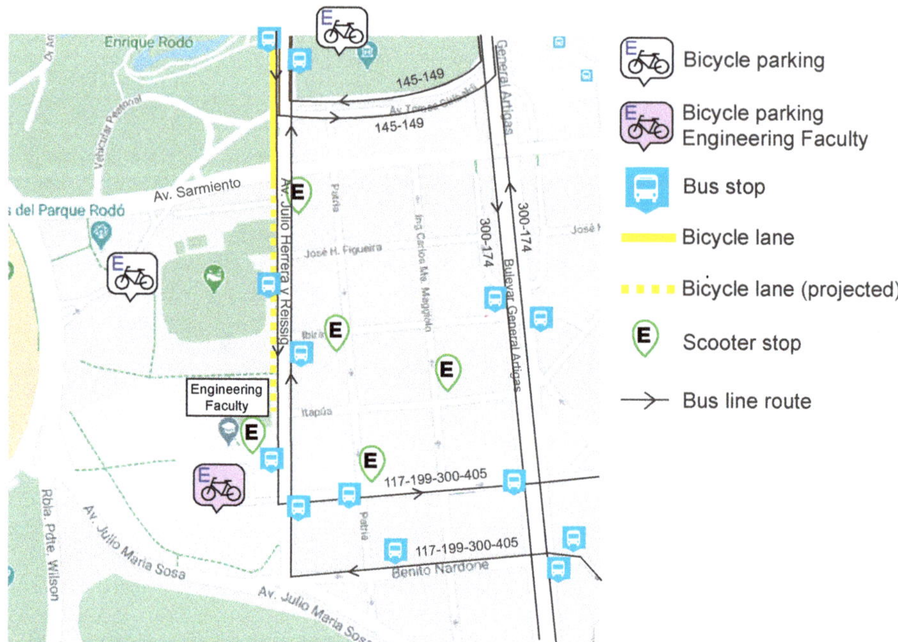

Figure 5. Coverage of Engineering Faculty and Parque Rodó neighborhood by the studied transportation modes.

Transportation modes. Regarding the transportation modes used by people commuting to/from Parque Rodó and Engineering Faculty, Table 5 reports the number of trips using each transportation mode declared in the survey and the percentage that it represents over the total. Transportation modes are listed from more sustainable to less sustainable.

Table 5. Transportation modes used for commuting to/from Parque Rodó neighborhood and Engineering Faculty.

Transportation Mode	Number of Trips	Percentage
walking	83	13.0%
bicycle	40	6.3%
scooter	0	0.0%
bus	361	56.4%
more than one transportation mode (on different days)	69	10.8%
non-sustainable transportation modes (car, motorcycle)	87	13.6%
total	640	100.0%

According to the results reported in Table 5, just 19.3% of the trips to/from Engineering Faculty and Parque Rodó are done using sustainable transportation modes. Overall, more than half of the trips are done using the bus. The number of trips using other non-sustainable transportation mode is 13% (mainly private cars, just 1.4% on motorcycle), almost the same than people walking to/from the studied area. Bus is the most popular transportation mode, mainly because it is the most accessible and affordable transportation mode for large distances, as confirmed by the accessibility and affordability analysis of transportation modes for the city on Montevideo, reported in Section 4.4.

Travel distances. A summary of distances traveled by people from/to Parque Rodó and Engineering Faculty is reported in the pie chart in Figure 6. Travel distances were calculated in Google Maps considering the origin the Engineering Faculty and the destination the neighborhood people reported.

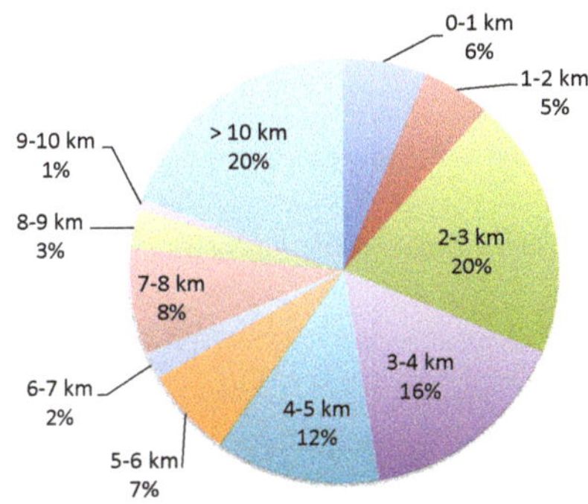

Figure 6. Travel distances according to data from the survey.

The analysis of travel distances indicates that 60% of the surveyed people commute from a maximum distance of 5 km away, and one-third of them travel between 2 to 3 km away. In addition, just 20% of the surveyed people commute a distance greater than 10 km. Furthermore, 95% of them declared to do a round trip, and 90% commute to the same place with a frequency of three times a week or more. These results confirms that the mobility demands in the studied zone follows a regular pattern, and that sporadic trips do not contribute significantly. Thus, the proposed approach, based on the analysis on frequent trips, provides a realistic characterization of mobility demands to/from Parque Rodó and Engineering Faculty.

Transportation modes by distance. Figure 7 refines the analysis of transportation modes, considering the average distance for each surveyed trip.

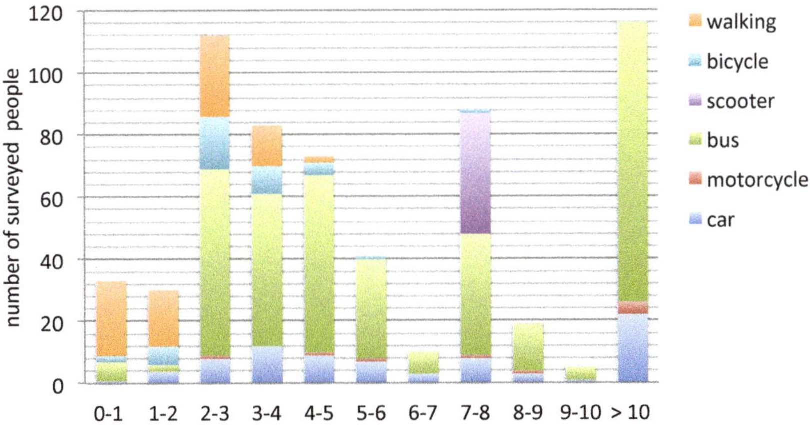

Figure 7. Transportation modes by distance.

The analysis of data reported in Figure 7 allows concluding that walking is the most popular transportation mode for distances less than 2 km, followed by bicycle and bus. For distances between 2 and 5 km, bus is the most popular transportation mode, followed by walking and bicycle. For distances longer than 5 km, bus is still the most used transportation mode, followed by non-sustainable transportation modes: car and motorcycle. Overall, the large number of people commuting to Parque Rodó and Engineering Faculty using non-sustainable transportation modes suggests that there is room to improve towards sustainable mobility in the studied area. Especially, 56.4% of trips using bus indicate that significant improvements to the service are definitely possible, by using electric buses. Specific actions can be also proposed to consider people traveling on car and motorcycles. This issue is studied in the following paragraphs, considering the information about preferences and motivations collected in the survey.

Commuting travel time. The combined analysis of distance and transportation modes allows computing the average commuting travel times for people commuting from/to the studied zone. In this regard, Table 6 reports the average travel times from/to the five most demanded origin/destination of surveyed trips, grouped by neighborhoods of Montevideo. The distance for each neighborhood is measured from Engineering Faculty to the centroid of each neighborhood.

Table 6. Commuting travel time to Engineering Faculty/Parque Rodó from the most frequent neighborhoods as origin/destination of trips.

Neighborhood	Distance	Bus	Bicycle	Scooter	Walking
Parque Rodó	1.0 km	-	4.4 min	7.0 min	12.0 min
Cordón	2.5 km	18.9 min	11.0 min	14.5 min	30.0 min
Tres Cruces	3.0 km	21.2 min	13.3 min	17.0 min	36.0 min
Pocitos	3.5 km	28.4 min	15.5 min	19.5 min	42.0 min
Centro	3.7 km	24.4 min	16.4 min	20.5 min	44.4 min
Prado	8.0 km	44.4 min	35.5 min	42.0 min	-

Results reported in Table 6 indicate that bicycle is the fastest transportation mode from distances shorter than 3 km. Scooter and bus are second and third regarding travel times, respectively. Considering that bicycles have no cost (either for using private vehicles or the public service projected for the zone) for up to 30 min, the bicycle is the fastest, most affordable, and most sustainable transportation mode for short distances. For distances between 3 km and 8 km, bicycle is the fastest transportation mode too, but a relevant issue that must be taken into account: when traveling long distances, commuters must consider that they might need to shower/change clothes due to the physical effort required, which would demand from a few to 10–15 additional minutes. Then, bus takes approximately the same time (i.e., 10 min more, but with no need to shower/change clothes) and scooter requires about four minutes more than bicycle, but considering that users can leave scooters in any place and there is no need to shower/change clothes, it is the fastest transportation mode. For distances larger than 8 km, all the studied transportation modes takes approximately the same time, so it is reasonable that most people use the bus, which is the most comfortable transportation mode, as reported in the previous analysis of transportation modes by distance.

Aspects people prioritize while commuting. Figure 8 summarizes the results of the analysis of those aspects identified as more relevant for people while commuting. The reported results are not necessarily linked to the transportation mode people use today, but to aspects they prioritize when commuting.

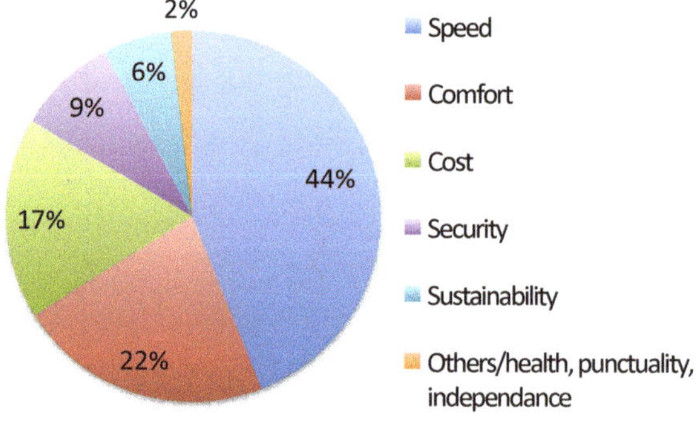

Figure 8. Aspects people prioritize while commuting.

The analysis of the aspects people prioritize while commuting indicates that 44% of the interviewed people prefer arriving faster to their destinations than other aspects. Comfort is the second feature more valued by the surveyed people (22%), and cost in third place (17%). Results obtained in the survey confirmed that aspects people prioritize do not depend on the distance or the travel time. In general, speed, comfort, and cost (in that order) are mentioned as priorities in declarations by surveyed people.

Overall, one conclusion can be formulated from the obtained results: public bus is the only transportation mode that could offer the three aspects people prioritize, in case it improves the actual service conditions. The other studied transportation modes cannot offer those three aspects, mainly because some of the are expensive (like private vehicles), ans several others require a physical effort and/or they are not comfortable in adverse climate conditions (like bicycle and scooter).

Willingness to change towards more sustainable transportation modes. The study interviewed 617 people, 504 of whom commute in non-sustainable transportation modes and 113 in sustainable transportation modes (bicycle or walking). Of those 504 people, 468 would be willing to change to sustainable transportation modes. This is a very relevant result, and is accounted as an empirical metric to determine the public acceptance of sustainable transportation modes within the people interviewed in our research. Furthermore, it is a first hint of the positive views towards sustainability, which can be confirmed by performing similar interviews in other (representative) neighborhoods of Montevideo. Figure 9 reports the results of the analysis of the sustainable transportation modes people would be willing to change.

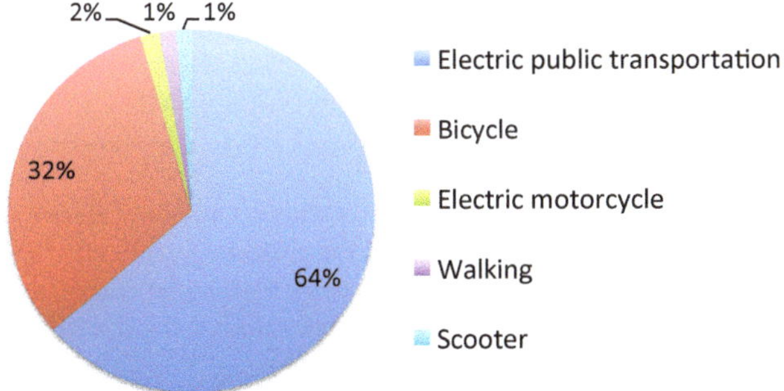

Figure 9. Sustainable transportation mode people who commute in non-sustainable transportation mode might change.

According to the results reported in Figure 9, electric public transportation is the mode that most of the people would be willing to change (64%), followed by bicycle (32%). Considering that CUTCSA and other bus companies plan to develop the modal shift from diesel to electric buses after evaluating the results of the pilot plan explained in Section 4, the willingness to change can provide a big leap in sustainable mobility in the studied area.

Taking into account the aforementioned results, the study analyzes next the reasons why people would like to change and why they do not actually change to both preferred transportation modes (electric bus and bicycle).

The reasons why people are willing to change to electric bus are mainly related to be part of initiatives oriented towards decarbonizing public transportation to reduce climate change and mitigate the environmental impacts of fossil fuels. Energy efficiency is also a motivation, especially considering that Uruguay is one of the leader countries in renewable energy in the world and it has a surplus of generated energy (over 98% of it generated from clean resources, according to reports for 2019 [47]). These opinions are in line with recognized benefits that electric public transportation provides to to the communities they serve (improving air quality, reducing greenhouse gas emissions, financial benefits related to reduced maintenance and operating costs, and avoiding healthcare expenses) [48].

On the other hand, Figure 10 summarizes some reasons why people might not change to electric public transportation. The analysis considers that no modifications on the current routes and frequencies will be associated to the electric bus, which will operate on the same conditions of the actual service (as suggested by the pilot plan implemented by CUTCSA).

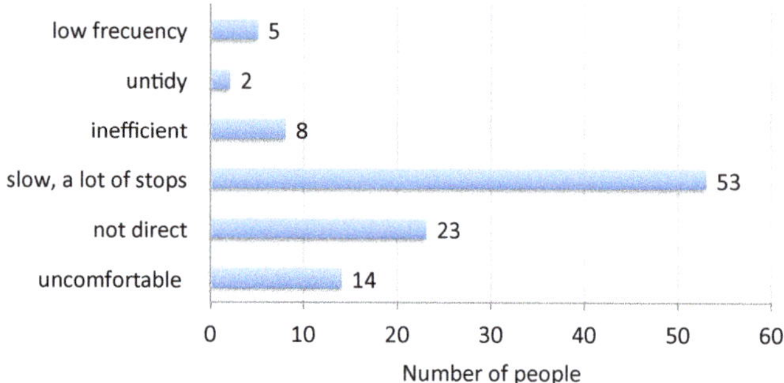

Figure 10. Reasons why people might not change to an electric public transport.

Regarding the results reported in Figure 10, the study collected opinions of 49 persons that travel in cars and would be willing to change to electric public transportation. However, most of them declared they will not change in case the bus will be inefficient, slow, untidy, with low frequency, and not direct. Additionally, 151 persons that travel today by bus would be willing to change to electric public transportation, even though most of them declared that they would also like to be faster, more comfortable, and more direct.

According to the survey, 149 persons reported that they would be willing to change to bicycle as transportation mode. In turn, 114 persons declared the reason why actually they do not use bicycle for commuting. The reasons why people are willing to do that modal change are related to the main benefits of riding a bicycle regarding health and also because it is the cheaper transportation mode, just as reported for the case of study of Montevideo.

On the other hand, Figure 11 summarizes the main reasons why people do not change their actual transportation mode to bicycle. Results correspond to 114 persons (eight traveling by car and 106 traveling by bus), who gave additional information about the reasons that prevent them to switch to bicycle. Car travelers mostly indicated the main reason is the reduced traffic safety, mainly because the few bike lines available in the city. In turn, bus travelers declared the main reasons for not using bicycle are the lack of proper facilities (e.g., their workplace does not offer a parking bicycle), they do not have enough space at home, and because they cannot afford a bicycle. Most actual bus travelers mentioned that they would use a public bicycle system if it was operating in the area.

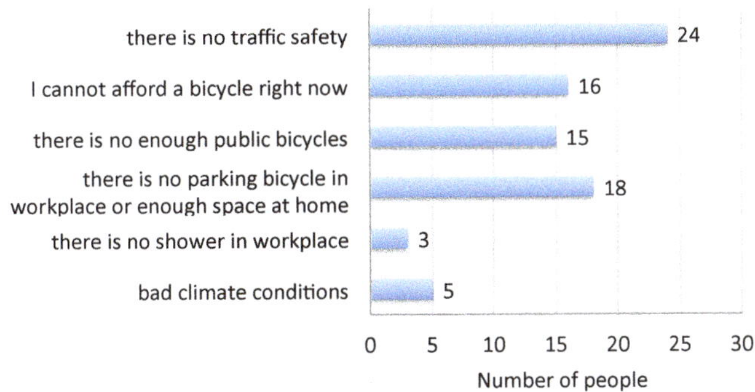

Figure 11. Reasons why people do not change to bicycle.

5.3. General Recommendations for Sustainable Mobility Initiatives in Parque Rodó Neighborhood and Engineering Faculty

This subsection provides specific suggestions and recommendations for improving sustainable mobility in the Engineering Faculty and Parque Rodó neighborhood. Recommendations and suggestions are based on the review, main results, and analysis of the mobility demands of the studied area, reported in previous subsections, especially considering the following concepts; the aspects related to accessibility (explored from the point of view of the infrastructures and services and and also from the point of view of people that commute from/to the studied area); the detected mobility patterns; and the motivation and opinions of interviewed people, which in fact constitutes a direct contribution of our research, as no previous similar studies have been developed in Montevideo.

One of the main facts observed from the analysis is that the studied area is fully covered by all the studied transportation modes (bus, bicycles, and electric scooters). This fact makes it easy connecting people with other zones of the city directly or via transfer trips. However, in terms of sustainable mobility, the neighborhood is not covered by the public bicycle initiative, which was analyzed for the case study of Montevideo, and the electric public transportation of the pilot plan of CUTCSA developed during 2016–2019 only performed 3.6% of the total trips traveled through the studied area. In this regard, one of the main recommendations is related to expanding the coverage area of public bicycles, by introducing bicycle stations in this area and design an articulated network of exclusive lanes, which also will help to improve other indicators, besides coverage.

Regarding the used transportation modes, more than half of the trips from/to the studied area are made by bus. For distances longer than 2 km, bus is the most popular transportation mode. Furthermore, 64% of the interviewed people that travel using non-sustainable transportation modes would be willing to change to electric public transportation. In addition, results of the study confirmed that people will not change to electric public transportation if the conditions of the service remain as nowadays. This is an important result because the pilot plan implemented by CUTCSA was developed in identical conditions than the actual service, regarding routes, bus stops, travel times, and other relevant indicators. In this regard, several suggestions are related to improve public transportation, in order to provide more useful and efficient sustainable transportation systems.

According to the commuting travel time indicator, some people declared they are not willing to change their actual transportation mode for electric public transportation because buses are very slow, they have many stops, low frequency, and routes are not direct. Thus, some suggestions for the new electric public transportation relate to introducing lines with fewer stops and higher frequencies than the current service to allow commuters, especially those whose trips demand more than 50 min (10 km), arrive faster to destination.

In terms of comfort and pleasure, bus companies can offer a better quality of service by improving the comfort of vehicles. Some ideas to give users a better service in terms of comfort and pleasure include improving travel conditions (e.g., appropriate space, air-conditioning, and free WiFi), guaranteeing universal accessibility, providing accurate real-time information via mobile applications and digital screens in bus stops, reducing motor vibration and noise, among others.

Regarding infrastructure, in 2010 Montevideo incorporated bus-only lanes in main roads and avenues, to avoid traffic congestion and speed up public transportation. However, bus lines that circulate through the studied area still share the same lane with other transportation modes. In this regard, the studied area has few wide avenues to install bus-only lanes; e.g., Herrera y Reissig, which crosses Parque Rodó neighborhood, has only one line in each direction from Sarmiento to Sosa (end of the avenue), thus including a second (bus-only) line would require a major infrastructure modification. However, mobility can still benefit for installing bus-only lines in Bulevar Artigas or in Herrera y Reissig (north, where there is space available). Regarding infrastructure for bicycle, considering the surveyed responses, two relevant suggestions are formulated to foster the modal shift: (i) bicycle lines should be extended, at least to include the projected line that will reach Engineering Faculty (which is planned since 2013, and has not been constructed due to non-disclaimed reasons); (ii) in addition, companies located in the zone and also in the main destination neighborhoods should be encouraged to provide bicycles parking within workplaces and also restrooms with showers, to be used by employees after the physical effort required for a ride.

The reported results, descriptive statistics, and suggestions are very valuable for the city administration in order to conceive an effective sustainable mobility plan in the studied area.

Finally, we acknowledge the implicancies of the reported analysis on policies and decision-making related to two relevant research and development initiatives our research group is currently participating on: (i) local sustainable mobility plans, developed by Ministry of Industry, Energy, and Mining through the MOVES project, and (ii) project "Spatial, universal, and sustainable accessibility: characterizing the multimodal transport system of Montevideo, Uruguay", developed with the support of the local administration (Intendencia de Montevideo), with the main goal of creating valuable knowledge and formulate specific policies to develop and improve mobility and accessibility. Some specific examples that can benefit from the analysis reported in the previous subsection for Parque Rodo neighborhood are the redesign of bicycle lines in the zone and the planning of a route for the electric bus to provide mobility services to Engineering Faculty and other faculties in the district (Architecture Faculty and Economics Faculty).

6. Conclusions and Future Work

This article studied sustainable mobility initiatives implemented in Montevideo, Uruguay, and a specific case study following a practical methodology to characterize and improve sustainable mobility in the Parque Rodó neighborhood and Engineering Faculty.

The study analyzed the main concepts of sustainable mobility by a review of related work on the topic and applied the existing knowledge to analyze three sustainable transportation modes currently available in Montevideo (electric bus, public bicycles, and electric scooters) through quantitative and qualitative indicators of sustainable mobility. Results of the study confirmed that the coverage area of the studied sustainable mobility initiatives is a small fraction of the total area of the city, thus a significant part of the population of Montevideo cannot access to sustainable transportation modes. Regarding cost, public bicycle is the most affordable mode of transportation, and electric bus is the second best option, mainly because these two services benefit from subsidies and support from public finances, thus they can keep a reasonable price for users. Electric scooters have prohibitive prices for low-income citizens. Public bicycle is also the fastest and the most ecological option for short and long distance travels. On the other hand, the quality of service of the public bicycle system, regarding comfort and pleasure, is the worst of the three studied transportation modes. Public bicycle, electric bus, and electric scooter provide intermodal connectivity between them, but there is a lack of intermodal integration between services. Specific suggestions were provided in regard of the main drawbacks of current sustainable mobility initiatives in Montevideo.

The mobility analysis of Parque Rodó neighborhood and Engineering Faculty was based on a survey performed to 617 persons who commutes to/from the neighborhood from/to other zones of the city. This is an important contribution of the reported research, since no previous analysis of sustainable mobility has been performed for specific zones of Montevideo.

Results of the study indicate that the area is fully covered by all the studied transportation modes (bus, bicycle, and electric scooters). However, in terms of sustainable mobility, the neighborhood lacks of a proper coverage, as the public bicycle initiative does not operate in the area and electric public transportation only did 3.6% of the total trips in 2016–2019 through the studied area. The survey reported that more than half of the trips from/to the studied area are made by bus. Bus is also the most popular transportation mode for distances longer than 2 km. This is a relevant result, because, according to the survey, more than half of the persons that currently travels using non-sustainable transportation modes would be willing to change to electric bus. Thus, improving the bus public transportation service towards the modal shift to electric emerges as a first priority.

On the other hand, despite the fact that the pilot plan for electric buses was successfully deployed and valued by citizens, and this transportation mode will provide users a better service and move towards sustainable mobility, people declared they are not willing to change their actual transportation mode if electric buses operate in the same conditions to the actual service. As a consequence, the shift from diesel to electric does not ensure the modal change to public sustainable transportation if buses still operate as in the actual service. To attract more users, bus companies should work on improving speed (one of the most important aspect people prioritize while commuting) by rethinking the routes and stops of new electric buses and also travel conditions in terms of comfort and pleasure (e.g., providing appropriate space, air-conditioning, free WiFi, etc.), which is another of the main aspects to solve to succeed in the modal change of citizens.

An important result of the mobility analysis in Parque Rodó and Engineering Faculty is that more than half of the surveyed trips involve distances shorter than 5 km, i.e., suitable distances to commute by bicycle. The study also reported that a quarter of the interviewed people are willing to switch to bicycle as a sustainable mobility alternative. However, the lack of infrastructure (bicycle lanes, bicycle parking and showers at workplace) discourages people to use the bicycle for commuting to/from the studied zone. Thus, improving the mobility conditions for bicycles turn to be an important topic to address in the studied zone.

Overall, the analysis of the three sustainable initiatives in Montevideo and the study of the sustainable mobility situation and demands of Parque Rodó neighborhood and Engineering Faculty are valuable tools for helping academics, transportation companies, and stakeholders to analyze and evaluate possible solutions to implement sustainable mobility plans. The reported case study in Parque Rodó provides a basis for building more powerful surveys and data collection activities to better understand sustainable mobility in the whole city.

The main lines for future work are related to extend the analysis, by including some studies that were not taken in consideration, e.g., discussion of automobility in the city (traffic volume data, number of traffic accidents, level of air pollution) and the measures that the city of Montevideo has taken to overcome the problems generated by the automobile, building specific origin–destination maps to account for the most demanded travels to/from Parque Rodó neighborhood and performing a deep study of the reasons why people would be willing to change to sustainable transportation modes and a deeper analysis of public acceptance via global interviews. The analysis of differential mobility needs is also an important line for future research, to assess the impact of sustainable mobility in Montevideo, focusing on its citizens. Regarding the replicability of the study, the proposed methodology can be applied to characterize the mobility demands and the sustainable mobility analysis on other relevant neighborhoods in the city. Furthermore, the analysis can be also extended by considering other sustainable mobility indicators and studying best practices implemented on other cities, in order to contribute to the improvement of urban sustainable mobility in Montevideo.

Author Contributions: Conceptualization, S.H. and S.N.; methodology, S.H. and S.N.; software, S.H. and S.N.; validation, S.H. and S.N.; formal analysis, S.N.; investigation, S.H. and S.N.; resources, S.H. and S.N.; data curation, S.H.; writing–original draft preparation, S.H. and S.N.; writing–review and editing, S.N.; visualization, S.H. and S.N.; supervision, S.N.; project administration, S.N. ; funding acquisition, S.H. and S.N. All authors have read and agreed to the published version of the manuscript.

Funding: This research received no external funding.

Conflicts of Interest: The authors declare no conflict of interest.

References

1. Harvey, D. Social Justice, Postmodernism and the City. *Int. J. Urban Reg. Res.* **1992**, *16*, 588–601. [CrossRef]
2. Cardozo, O.; Rey, C. La vulnerabilidad en la movilidad urbana: Aportes teóricos y metodológicos. In *Aportes Conceptuales y Empíricos de la Vulnerabilidad Global*; Foschiatti, A., Ed.; Editorial Universitaria de la Universidad Nacional del Nordeste: Chaco, Argentina, 2007; pp. 398–423.
3. Jeon, C.; Amekudzi, M. Addressing Sustainability in Transportation Systems: Definitions, Indicators, and Metrics. *J. Infrastruct. Syst.* **2005**, *11*, 31–50. [CrossRef]
4. Barrionuevo, J.; Berrone, P.; Ricart, J. Smart cities, sustainable progress. *IESE Insight* **2012**, *14*, 50–57. [CrossRef]
5. Miller, P.; de Barros, A.; Kattan, L.; Wirasinghe, S. Public transportation and sustainability: A review. *KSCE J. Civ. Eng.* **2016**, *20*, 1076–1083. [CrossRef]
6. Ministry of Industry, Energy, and Mining, Uruguay. Proyecto MOVÉS: Movilidad Urbana Eficiente y Sostenible. 2017. Available online: https://www.miem.gub.uy/energia/proyecto-moves-movilidad-urbana-eficiente-y-sostenible (accessed on 1 August, 2019).
7. Hipogrosso, S.; Nesmachnow, S. Sustainable Mobility in the Public Transportation of Montevideo, Uruguay. In *Smart Cities*; Communications in Computer and Information Science; Springer: Berlin/Heidelberg, Germany, 2020; Volume 1152, pp. 93–108.
8. General Assembly of the United Nations. Report of the World Commission on Environment and Development: Our Common Future. 1987. Available online: http://www.ask-force.org/web/Sustainability/Brundtland-Our-Common-Future-1987-2008.pdf (accessed on 1 August 2019)
9. World Business Council for Sustainable Development. Methodology and Indicator Calculation Method for Sustainable Urban Mobility. Technical Report 978-2-940521-26-5, WBCSD, 2015. Available online: https://www.eltis.org/sites/default/files/trainingmaterials/smp2.0_sustainable-mobility-indicators_2ndedition.pdf (accessed on 1 May 2019)
10. General Assembly of the United Nations. Sustainable Development Goals. 2015. Available online: https://sustainabledevelopment.un.org/ (accessed on 1 April 2020).
11. Banister, D. The sustainable mobility paradigm. *Transp. Policy* **2008**, *15*, 73–80. [CrossRef]
12. Marshall, S. The challenge of sustainable transport. In *Planning for a Sustainable Future*; Layard, A.; Davoudi, S.; Batty, S., Eds.; Spon: London, UK, 2001; pp. 131–147.
13. C40 Cities Climate Leadership Group. Good Practice Guide. Transport Oriented Development. 2016. Available online: https://www.c40.org/networks/transit-oriented-development (accessed on 1 May 2020).
14. Gudmundsson, H.; Hall, R.; Marsden, G.; Zietsman, J. *Sustainable Transportation*; Springer: Berlin/Heidelberg, Germany, 2016.
15. Xiong, Z.; Sheng, H.; Rong, W.; Cooper, D. Intelligent transportation systems for smart cities: A progress review. *Sci. China Inf. Sci.* **2012**, *55*, 2908–2914. [CrossRef]
16. Litman, T.; Burwell, D. Issues in sustainable transportation. *Int. J. Glob. Environ. Issues* **2006**, *6*, 331–347. [CrossRef]
17. Litman, T. Exploring the Paradigm Shifts Needed To Reconcile Transportation and Sustainability Objectives. *Transp. Res. Rec.* **1999**, *1670*, 8–12. [CrossRef]
18. Rodrigues, A.; Costa, M.; Macedo, M. Multiple views of sustainable urban mobility: The case of Brazil. *Transp. Policy* **2008**, *15*, 350–360.
19. Johnston, R. Indicators for Sustainable Transportation Planning. *Transp. Res. Rec. J. Transp. Res. Board* **2008**, *2067*, 146–154. [CrossRef]

20. Maclaren, V. Exploring the paradigm shift needed to reconcile sustainability and transportation objectives. *J. Am. Plan. Assoc.* **1999**, *62*, 184–202. [CrossRef]
21. Rodrigues, A.; da Silva, M.; Rodrigues, R. Development and application of I_SUM—An index of sustainable urban mobility. Transportation Research Board Annual Meeting. 2010. Available online: https://www.researchgate.net/publication/277124018_Development_and_application_of_I_SUM_an_index_of_sustainable_urban_mobility (accessed on 4 May 2019)
22. Baidan, A. A Brief Analysis of the Sustainable Mobility Approach in Bucharest. *Procedia Environ. Sci.* **2016**, *32*, 168–176.
23. Lyons, W. Sustainable Transport in the Developing World: A Case Study of Bogota's Mobility Strategy. In Proceedings of the International Conference on Sustainable Infrastructure, New York, NY, USA, 26 October 2017.
24. Nesmachnow, S.; Tchernykh, A.; Cristóbal, A. Planificación de transporte urbano en ciudades inteligentes. I Ibero-american Conference on Smart Cities, Soria, Spain, 26–27 September 2018; pp. 204–218. Available online: https://www.fing.edu.uy/eventos/WPTCI/presentaciones/PlanificaciC3B3ndetransporteurbanoenciudadesinteligentes.pdf (accessed on 1 May 2019)
25. Ministry of Industry, Energy, and Mining, Uruguay. Plan Institucional de Movilidad Sostenible. 2019. Available online: https://moves.gub.uy/iniciativa/plan-institucional-de-movilidad-sostenible (accessed on 1 March 2020).
26. Abreu, P.; Vespa, J. Plan de Movilidad. 2010. Available online: http://www.montevideo.gub.uy/sites/default/files/plan_de_movilidad.pdf (accessed on 30 December 2018).
27. Acosta, I. Uruguay's Public Transport Goes Electric. Tierramerica. Environment and Development. Available online: http://www.ipsnews.net/2014/03/uruguays-public-transport-goes-electric (accessed on 24 March 2014).
28. CUTCSA. Reporte Social 2016-2017. Available online: http://www.cutcsa.com.uy/content/uploads/2019/02/Reporte-Social-2016-2017.pdf (accessed on 1 August 2019).
29. Infonegocio. El Estado subsidiará para que haya 120 ómnibus totalmente eléctricos y sin escalones. Available online: https://infonegocios.biz/enfoque/el-estado-subsidiara-para-que-haya-120-omnibus-totalmente-electricos-y-sin-escalones (accessed on 1 January 2020).
30. Presidencia de la República. Subsidios de Gobierno e Intendencia posibilitan que 54 taxis eléctricos funcionen en Montevideo. December 2018. Available online: https://www.presidencia.gub.uy/comunicacion/comunicacionnoticias/54-taxis-electricos-miem-ute-montevideo (accessed on 20 January 2020).
31. Intendencia de Montevideo. Ciudad Vieja a Escala Humana. 2015. Available online: https://montevideo.gub.uy/ciudad-vieja-a-escala-humana (accessed on 1 May 2020).
32. Shen, L.; Du, L.; Yang, X.; Du, X.; Wang, J.; Hao, J. Sustainable Strategies for Transportation Development in Emerging Cities in China: A Simulation Approach. *Sustainability* **2018**, *10*, 844. [CrossRef]
33. Toutouh, J.; Lebrusán, I.; Nesmachnow, S. Computational Intelligence for Evaluating the Air Quality in the Center of Madrid, Spain. In *Communications in Computer and Information Science*; Springer: Berlin/Heidelberg, Germany, 2020; pp. 115–127.
34. Anagnostopoulou, E.; Bothos, E.; Magoutas, B.; Schrammel, J.; Mentzas, G. Persuasive Technologies for Sustainable Mobility: State of the Art and Emerging Trends. *Sustainability* **2018**, *10*, 2128. [CrossRef]
35. Atash, F. Redesigning Suburbia for Walking and Transit: Emerging Concepts. *J. Urban Plan. Dev.* **1994**, *120*, 48–57. [CrossRef]
36. Carruthers, R. Affordability of Public Transport. In *Competition and Ownership in Land Passenger Transport: Selected Papers from the 9th International Conference*; Emerald Group Publishing Limited: Bingley, UK, 2005; pp. 1–15.
37. Instituto Nacional de Estadística, Uruguay. Ingresos de los hogares y de las personas. Available online: https://www.ine.gub.uy/gastos-e-ingresos-de-las-personas-y-los-hogares (accessed on 15 August 2019).
38. Mauttone, A.; Hernández, D. Encuesta de movilidad del área metropolitana de Montevideo. 2017. Available online: https://scioteca.caf.com/bitstream/handle/123456789/1078/EncuestadeMovilidadMVD-documentocompleto-final.pdf (accessed on 15 August 2019).
39. Nesmachnow, S.; Bana, S.; Massobrio, R. A distributed platform for big data analysis in smart cities: combining intelligent transportation systems and socioeconomic data for Montevideo, Uruguay. *EAI Endorsed Trans. Smart Cities* **2017**, *2*, 1–18. [CrossRef]

40. Massobrio, R.; Nesmachnow, S. Urban data analysis for the public transportation systems of Montevideo, Uruguay. In Proceedings of the 2nd Iberoamerican Congress on Smart Cities, Soria, Spain, 7 October 2019.
41. Statistics Canada. Transportation. 2020. Available online: https://www150.statcan.gc.ca/n1/pub/11-402-x/2012000/chap/trans/trans-eng.htm (accessed on 1 March 2020).
42. Bureau of Transportation Statistics, USA. Daily Passenger Travel. Available online: https://www.bts.gov/archive/publications/highlights_of_the_2001_national_household_travel_survey/ (accessed on 1 March 2020).
43. Hollingsworth, J.; Copeland, B.; Johnson, J.X. Are scooters polluters? The environmental impact of shareddockless electric scooters. *Environ. Res. Lett.* **2019**, *14*, 1–10. [CrossRef]
44. El País. IMM Licita 60 Estaciones Para 600 Bicicletas en Montevideo. Available online: https://www.elpais.com.uy/informacion/sociedad/licitan-bases-bicicletas-montevideo.html (accessed on 24 June 2019).
45. Fernandez, G.; Fernandez, A. Strategic Thinking for Sustainability: A Review of 10 Strategies for Sustainable Mobility by Bus for Cities. *Sustainability* **2018**, *10*, 4282. [CrossRef]
46. Universidad de la República. Basic Statistics 2018. Available online: gestion.udelar.edu.uy/planeamiento/ (accessed on 1 March 2020).
47. International Energy Agency. World Energy Outlook 2019. Available online: https://www.iea.org/reports/world-energy-outlook-2019 (accessed on 21 December 2019).
48. Horrox, J.; Casale, M. Electric Buses in America: Lessons from Cities Pioneering Clean Transportation. Available online: https://uspirg.org/sites/pirg/files/reports/ElectricBusesInAmerica/US_Electric_bus_scrn.pdf (accessed on 22 October 2019).

Article

Using Smart City Tools to Evaluate the Effectiveness of a Low Emissions Zone in Spain: Madrid Central

Irene Lebrusán [1,†] and Jamal Toutouh [2,*,†]

1 Institute for Global Law and Policy, Harvard University, Cambridge, MA 02138, USA; ilebrusan@law.harvard.edu
2 Computer Science and Artificial Intelligence Laboratory, Massachusetts Institute of Technology, Cambridge, MA 02139, USA
* Correspondence: toutouh@mit.edu
† These authors contributed equally to this work.

Received: 4 May 2020; Accepted: 19 May 2020; Published: 1 June 2020

Abstract: Population concentration in cities brings new risks as an increase in pollution, which causes urban health problems. In order to address this problem, traffic reduction measures are being implemented as pedestrianization areas; they are the definition of Low Emissions Zones (LEZs). When the effectiveness of these types of measures is in doubt, smart city tools provide data that can be used to scientifically asses their impact. This article analyzes the situation of Madrid Central (Spain), a LEZ subject to controversy. We apply statistical and regression analyses to evaluate the effectiveness of this measure to reduce air pollution and outdoor noise. According to the results, this LEZ was able to significantly reduce NO_2, $PM_{2.5}$, and PM_{10} concentration locally, having the same positive impact in the rest of the city. In terms of noise, this measure was able to mitigate background noise levels generated by road traffic.

Keywords: air pollution; noise pollution; low emissions zone; pedestrianization

---

1. Introduction

Urbanization trends worldwide show a clear preference of the population to live in urban areas: Today, 55% of the world's population lives in urban places and this proportion is expected to increase to 68% by 2050 [1]. The attraction of the city lies in the numerous opportunities resulting from the agglomeration of human capital and the associated economic dynamism. However, while it is true that urban areas are key for social change, wealth generation, and innovation, the concentration of people in cities also leads to new risks from an economic, social, and environmental point of view.

It is not uncommon for cities to be car-use oriented, with a design characterized by lack of safe public spaces that tend to discourage physical activity [2]. Nonetheless, the major concern derived from the rapid development of car-oriented cities is the high generation of emissions (air pollutants and noise) and their impact on the inhabitants' health [3]. Another related problem is the increase of environmental noise, which gives rise to a range of health problems including annoyance, sleep disturbance, increasing hypertension, and cardiovascular diseases [4]. In this regard, the speed of urbanization has outpaced the ability of governments to build essential infrastructures that make life in cities safe, rewarding, and healthy [5].

As about 80% of this pollution associated with health problems is produced by cars [6], reducing traffic seems to be an efficient strategy to improve urban livability and inhabitants' health. Thus, many cities around the world have started to shift toward non-car friendly access (pedestrianization measures) because it is an effective and non-expensive way to reduce air pollution and noise [3,7–10].

According to the European Commission, a smart city goes beyond the use of information and communication technologies (ICT) for better resource use and less emissions [11]. In fact, different smart city tools and smart city management can be invaluable for fulfilling the government's commitment to the citizen. Taking advantage of this approach and, in order to measure effectiveness or even to scientifically guide the process, we can benefit from putting into use air quality monitoring and other tools that a smart city provides, e.g., internet of things (IoT) and open data.

In this work we analyze with smart city tools the effectiveness of the implementation of Madrid Central (MC), a Low Emissions Zone (LEZ) in Madrid, Spain. This approach is specifically needed there because of the controversy around MC and its use as a political symbol. Despite the fact that this LEZ was actually an extension and continuation of previous measures with the goal to comply with a European air quality requirement, MC encountered strong opposition. Those claiming the supposed right to go by car claimed this measure is ineffective concerning the reduction of pollutants and claim it is a cause of border effect (pollution displacement). Fulfilling its electoral promise, after the elections the new government decided to reverse the MC measures.

This article is an extension of a preliminary study in which this pedestrianization measure was locally evaluated in terms of its impact on the LEZ on a subset of pollutants [12]. The data used in that study comprises a time frame of 30 months (between December 2016 and May 2019), even if MC was active just during six months. According to the results obtained, the research concluded that MC was effective in reducing the concentration of NO_2 and noise in the area.

This new analysis considers a longer amount of time and a more comprehensive range of pollutants. This piece also takes into consideration a bigger area of the city in order to check potential side effects. Therefore, in this article, we face the following research questions: RQ1: Is the deployment of MC effective at reducing the concentration of NO_2 and so improving the air quality in this area? RQ2: Is the implementation of an LEZ an effective measure to reduce environmental noise levels?. RQ3: Do pedestrianization policies in a given area of the city produce a pollution displacement to other zones of the city?. RQ4: Are smart city tools effective for evaluating urban health policies and other measures implemented in the city?.

The main contributions of this study are: (i) It evaluates the effectiveness of traffic limitation measures in reducing emissions harmful to health; (ii) it analyzes the impact of such measures on environmental noise, which is also associated with poorer urban health; (iii) it checks whether the claim of pollution displacement (border effect) has any basis according to results; and (iv) it appraises the usefulness of smart city tools as an appliance for evaluating urban measures and policies to reduce pollution.

The paper is organized as follows: First, the impact of pollutants on health is reviewed, MC is contextualized, and related research is introduced. Section 3 introduces the materials and methods used in this analysis. Air quality and noise reduction are evaluated first locally (in the LEZ) and in the whole city in Sections 4 and 5. Finally, Section 6 presents the conclusions and the main lines of future work.

2. Background

This section introduces first the main effects of car-use oriented cities on health. Next, it presents the EU directives on air pollution and the main aspects of the implementation of MC, including the municipal regulations on which it is based. Finally, it reviews relevant related work.

2.1. Are We Designing Cities for Cars or for People? The Effects on Health and Well-Being

The urban design of most cities prioritizes the use of motorized vehicles, thus relegating pedestrian travel. This approach to the city has a variety of negative impacts on safety and health, as well as on the general quality of life. An urban design focused on car use has a negative effect on the quality of the social space, forgetting that the physical and spatial context has a deep influence on people throughout the whole course of life. It has been proved how the growth of car use reduces children's access to public space in urban contexts due to traffic and lack of infrastructure for pedestrians [13,14]. Restricting mobility is critical for the development of children's spatial awareness and spatial activity, thus affecting their social and physical development [15]. Other groups, such as the elderly, could benefit too from environments with safe walking access, improving their independence and well-being [16].

Moreover, traffic and car use cause a high generation of emissions, both air pollutants and noise. This leads to a range of urban health problems: It reduces life expectancy, produces loss of years of a healthy life, and diminishes the quality of health [3]. Air pollution, which caused more than 400,000 premature deaths in 2016, is considered the top health hazard in the European Union (EU) [17,18]. It has proved to be associated with heart disease and strokes, lung diseases, and lung cancer, besides reducing lung capacity and aggravating asthma. Air pollution has been also pointed out as carcinogenic and causes infertility and diabetes type 2 [19]. Other studies link air pollution to obesity, systemic inflammation, ageing, Alzheimer's disease, and dementia [20]. It also affects the brain in the same way that Alzheimer's does as it causes changes in the structure of the brain [21]. While pollution affects all ages, some population groups are more vulnerable to pollution problems, such as pregnant women, newborns, children, and the elderly. It also can exacerbate preexisting conditions at all ages.

Environmental noise is the major preventable cause of hearing loss [4]. It can also cause a range of non-auditory problems. To begin with, the evidence for the effects of environmental noise on health is strongest for annoyance, sleep disturbance, and cognitive performance in both adults and children [22]. Sound pollution also affects the cardiovascular system and causes hypertension [4,22]. Among children, it generates cognition problems, such as communication difficulties, impaired attention, increased arousal, frustration, and worse performance [23,24]. Last but not least, noise pollution causes annually at least 16,600 cases of premature death in Europe [25].

These risks could be reduced by limiting car use. Among different approaches to reducing car use, pedestrianization and LEZ have been proved as some of the most effective strategies against emissions. Absolute pedestrianization is difficult to implement (and expensive) but LEZs have proved to be successful in different European Member States. These are the cases for the LEZ ("Umweltzone") for trucks and cars in the center of Berlin inside the S-Bahn ring (Germany); the LEZ in the Lombardy Region for motorcycles, buses (whole year), and vehicles during wintertime, e.g., in Milan (Italy); or the LEZ for vans and lorries in Greater London (United Kingdom). All of these experiences have registered good results in terms of improving air quality.

2.2. Reducing Pollution: International Directives and the Low Emissions Zone in Madrid

While cities are related to social and economic progress, the increase of air pollution and noise can be considered a modern plague [26] decreasing the quality of life in cities [27]. In Europe (where 74% of the population lives in urban areas) the concern on this matter has led to the creation of a common set of environmental rules. These directives could not just safeguard EU citizens from environment-related pressures and risks to health and well being, but also reduce significantly different expenses: An adequate implementation of environmental legislation could save 50 billion euro every year in health and environment costs to the EU economy [20].

Because of this, a Clean Air Policy Package is adopted in 2013 by the the European Commission. This air quality policy, based on Directives 2008/50/EC [28] and 2004/107/EC [29] rests on three pillars: (i) Air quality standards; (ii) national emission reduction targets established in the National Emissions Ceiling Directive; and (iii) emissions standards for key sources of pollution (such as cars).

Noise being another harmful hazard, the EU regulation (Directive 2002/49/EC) [30], for every Member State: (i) Establishes the creation of noise mapping to determinate exposure levels to environmental noise; (ii) makes environmental noise information available to the citizens; (iii) establishes adoption of action plans, based upon noise-mapping results, to prevent and reduce environmental noise where necessary.

The referred to directive points out that road traffic is a majors source of noise; it has also been noted as the largest contributor to nitrogen oxide emissions [31]. For the reasons alluded to previously, NO_2 is one of the main concerns of the EU: Exposure to air pollution from road transport costs about 137 billion per year in Europe [32] but, more important, it produced around 76,000 premature deaths in 2015 [33]. Despite this, a number of countries keep exceeding the maximum established PM_{10} and NO_2 levels. Spain is one of them, with several urban areas under surveillance by the EU because of their poor air quality. The European Commission required the reduction of these air pollutants under the threat of penalties.

Madrid City Council decided to reduce transit traffic in a delimited area of the downtown, replicating other European experiences. Madrid Central is an LEZ in Madrid, consisting of a number of car access restrictions for non-residents, independent from the current pollution level of the air.

The clarity in the perimeter of MC is one of their virtues, so facilitating the understanding of zonal delimitation and helping to introduce a behavioral change in the city.

MC is created by the Ordenanza de Movilidad Sostenible (5 October 2018) and it covers almost the entire Centro district, which is formed by the neighborhoods of Palacio, Embajadores, Cortes, Justicia, Universidad, and Sol. In this area, of 4,720,000 m^2, Centro contains 134,881 people, of which 21.6% are over 65 years old and 9.2% are less than 17 years old. Children and the the elderly are the age groups which can benefit more from the reduction of noise and pollutants.

The goal of MC is to improve air quality, but also to respond to the idea of changing the uses of spaces in the city center, prioritizing the pedestrian one and reducing noise pollution. However, as we said, its conformation mainly responds to ensuring the objectives demanded by the EU. The traffic restriction started on 30 November 2018, and the fines for noncompliance did not start until 16 March 2019. Thanks to this straightforward measure, Spain avoided being brought to the European Court of Justice and so, the economic fine.

However, the measure raised strong opposition from some political parties. After the elections (held on 26 May 2019) and qualifying the MC measures as inefficient and even unnecessary, the newly elected government approved article 247 of the Ordenanza de Movilidad Sostenible, applying with it a moratorium on fines from 1 July to 30 September 2019. Besides a warning from the EU, the decision raised social protests, especially from environmental groups. This suspension led to the emergence of social movements claiming the paralysis of this reversal based on the negative effects on health and environment, and a warning from the EU. Finally, after a contentious administrative appeal filed by the Platform in Defense of Central Madrid, a judge reactivated MC, suspending the moratorium on fines.

2.3. Related Works

The proliferation of pedestrianization measures and policies in different cities has led to a number of studies to evaluate their efficacy. A brief review of the related literature is presented next.

The impact of the rapid growth of car ownership in Beijing, China was analyzed in terms of transportation, energy efficiency, and environmental pollution [34,35] A set of measures were applied in Beijing in 2010 to mitigate the effects of traffic congestion and reduce air pollution. Liguang et al. [34] analyzed data from Beijing Municipal Committee of Transport to evaluate the implementation of car use restriction measures. Fairly good effects on improving urban transportation and air quality were achieved according to the results reported. Liu et al. [35] proposed an indirect approach to evaluate the impact of car restrictions and air quality, by applying a generalized additive model to explore the association of driving restrictions and daily hospital admissions for respiratory diseases. Several interesting facts were obtained from the analysis, including higher daily hospital admissions for respiratory disease for some days, and the stronger effect on cold season. Females and people older than 65 years benefited more from the applied environmental policy. Overall, the authors found positive effects on the improvement of public health. More research had been performed addressing the LEZ analysis in China [36–38].

In Europe, more than 200 LEZ have been deployed in recent years [39]. Studies on their impact on urban air quality have been performed in several countries: Netherlands [40,41], Denmark [42], United Kingdom [43,44], Germany [45,46], Italy [47], and Portugal [48,49]. Most of them analyzed and acknowledged the reduction of two hazardous pollutants: Nitrogen dioxide (NO_2) and particulate matter (PM).

Regarding noise pollution, there is a body of work on analyzing urban noise and its impact on the population's health [50–55]. There are only a few studies done in this regard [12,56]. However, most of the studies presented above studied also some variables related to noise pollution.

Focusing on Madrid, air quality has been a health issue over recent decades. Thus, there are several studies dealing with air pollution in this city. Borge et al. [57] reported the modeling activities carried out in Madrid, emphasizing the atmospheric emission inventory development, which comprises the combination of models and real data. They showed that Madrid required a reduction in NO_x emissions in order to meet the NO_2 European standards, which was the main motivation behind the implementation of MC. Different models were used to simulate and evaluate a short-term action plan to mitigate pollution emissions [58,59]. More recently, after the application of MC, Lebrusán and Toutouh [12] evaluated the effectiveness of this measure in reducing pollution inside the LEZ. Another study applied computational intelligence (deep neural networks and regression analysis) to evaluate the evolution of NO_2 concentration in the air, but again only locally (inside the MC area) [60]. Both reported a decrease in NO_2 concentration. Our article contributes to this line of research by extending the set of pollutants studied by adding PM, among others, taking into account a longer time frame, and including 24 areas of study in order to get a more general idea of the impact of this type of measures in the whole of Madrid.

3. Materials and Methods

The primary goal of this study is to take advantage of smart city tools, such as a sensor network and open data, to evaluate the effects of an LEZ on environmental pollution and noise. We address such an analysis, first locally, evaluating the impact of this measure at the MC area, and globally, by assessing the same in different areas of the city (Madrid).

Madrid City Council installed different sensors (see Figure 1) that gather data on the ambient concentration of different pollutants and measure the output noise levels. The data is available on the Open Data Portal (ODP) Madrid City Council (URL: https://datos.madrid.es/). We use such open data in this research.

Sensor id.	Sensor location
4	Pza. de España
8	Escuelas Aguirre
11	Avda. Ramón y Cajal
16	Arturo Soria
17	Villaverde
18	Farolillo
24	Casa de Campo
35	Pza. del Carmen
36	Moratalaz
38	Cuatro Caminos
39	Barrio del Pilar
40	Vallecas
47	Mendez Alvaro
48	Castellana
49	Parque del Retiro
50	Plaza Castilla
54	Ensanche de Vallecas
55	Urb. Embajada
56	Pza. Elíptica
57	Sanchinarro
58	El Pardo
60	Tres Olivos

Figure 1. Location of the sensors that gather the pollution information shared through the ODP. The shaded area represents the Low Emissions Zone (LEZ) analyzed in this article.

The analysis is performed considering a temporal frame of six years (72 months), from December 2013 to November 2019. Two periods are distinguished: Pre-MC, i.e., the period of five years before implementing MC (from December 2013 to November 2018), and Post-MC, i.e., the period of one year after implementing the mobility measure (from December 2018 to November 2019). The main idea is to compare both periods to asses the effect of the measures applied in MC.

In the following, we introduce the air pollutants analyzed, the outdoor noise metrics studied, and the methodology applied in the evaluation.

3.1. Air Quality Evaluation

The Open Data Portal (OPD) provides the hourly mean concentration of several air pollutants. In this study we focus on six: Sulfur dioxide (SO_2), nitrogen dioxide (NO_2), ozone (O_3), carbon monoxide (CO), particulate matter 10 micrometers or less in diameter (PM_{10}), and particulate matter 2.5 micrometers or less in diameter ($PM_{2.5}$). Table 1 summarizes the maximum concentration allowed in the studied pollutants established by WHO and EU directives.

Table 1. Maximum concentration allowed of pollutants in the air by the WHO and EU regulations.

Pollutant	Averaging Period	WHO Theshold Concentration	EU Theshold Concentration
SO_2	24 h	20 µg/m^3	125 µg/m^3
	1 h	-	350 µg/m^3
	10 min	500 µg/m^3	-
NO_2	1 year	40 µg/m^3	40 µg/m^3
	1 h	200 µg/m^3	200 µg/m^3
O_3	8 h	100 µg/m^3	120 µg/m^3
CO	8 h	10 mg/m^3	10 mg/m^3
PM_{10}	24 h	50 µg/m^3	50 µg/m^3
	1 year	20 µg/m^3	40 µg/m^3
$PM_{2.5}$	24 h	25 µg/m^3	25 µg/m^3
	1 year	10 µg/m^3	-

3.2. Noise Evaluation

It is not trivial to chose the right parameters to evaluate noise pollution and its impact on the people. The sound-meters return measures that describe the physical attributes of the sound, but not the subjective response and the extent of the physiological and psycho-social harm done to the public. The noise pollution data provided by the OPD include the daily mean of the equivalent sound pressure levels, the percentile noise levels [61], and the noise pollution level (NPL) [62]:

- Equivalent sound pressure levels (L_{eq}) can be described as the average sound level for the measurement. The data analyzed include the L_{eq24} that corresponds to the noise measured during the whole day (24 h). In addition, three L_{eq} are returned by taking into account the period of the day: L_{eqD}, L_{eqE}, and L_{eqN}, which are evaluated during the day (from 7:00 h to 19:00 h), evening (from 19:00 h to 23:00 h), and night (from 23:00 h to 7:00 h), respectively.
- Percentile noise levels (L_x) are the levels exceeded for x percent of the time, where x is between 0.1% and 99.9%. L_x is calculated by applying statistical analysis. We evaluate the L_{10}, L_{50}, and L_{90}. The L_{10} and L_{90} are extensively used for rating any annoying traffic noise and background noise, respectively.
- NPL was developed to estimate the dissatisfaction caused by road traffic noise comprising the continuous noise level (L_{eq}) and the annoyance caused by fluctuations in that level. NPL is equal to L_{eq} plus 2.56 times the standard deviation of the noise distribution and it is generally approximated by Equation (1).

$$NPL \approx L_{eq} + (L_{10} - L_{90}) \tag{1}$$

All sound levels referred to in this article are measured in terms of an A-weighted decibel (dBA), which corresponds to the A-weighted sound level readings to replicate the response of the human ear to the annoyance caused by road traffic noise.

The Guideline Development Group (GDG) of WHO strongly recommends reducing L_{eq} noise levels produced by road traffic below 53 dBA. Road traffic noise above this level is associated with adverse health effects. Specifically, for night noise exposure (L_{eqN}), the GDG strongly recommends reducing noise levels produced by traffic flows below 45 dBA, since road traffic noise above this level is associated with adverse effects on sleep [63]. According to the Community of Madrid regulations presented in the Normativa de ruido diurno y vibraciones [64], the L_{eq} levels should be between 50 and 65 dBA during the daytime and between 40 and 55 dBA during the night.

3.3. Methodology

In order to asses the effectiveness of the LEZ (see Section 2.2), we evaluate the air pollution and the noise in the LEZ and in different areas of the city. Thus, we asses the local effect on the LEZ area and whether it is possible that a border effect is occurring as a result of potential traffic diversion (i.e., check the existence or absence of the border effect). The analysis performed in this article mainly considers:

- The pollutant concentration or level of noise itself during both periods Pre-MC and Post-MC, which is measured and averaged for periods of one hour and one day.
- The average difference between the pollutant concentration or level of noise x during Post-MC ($x^m_{Post-MC}$) and Pre-MC (x^m_{Pre-MC}), taking into account different months (M). We denote this metric by Δ (see Equation (2)). Negative values of Δ denote reduction/improvement of x.
- The percentage of the time the population is exposed to air pollutant concentration or noise levels below the thresholds defined by EU denoted by τ. These thresholds are shown in Table 1. The value of τ allows the assessment of the effectiveness of MC to potentially improve urban health because there may be situations where the pollution or noise is reduced but the situation is still unhealthy according to the EU regulations (e.g., NO_2 concentration $> 40\ \mu g/m^3$).
- Polynomial regression is applied to evaluate the general trend in air pollution concentration or levels of noise with and without the implementation of the road traffic restrictions in Madrid Central. While it is one of the simplest methods for analysis and estimation of time series, it is frequently used in the related literature [12,35,38] In this article, two polynomial regression methods are studied: Linear and polynomial of grade 10.

$$\Delta = \frac{1}{|M|} \sum_{m \in M} x^m_{Pre-MC} - x^m_{Pre-MC} \tag{2}$$

In order to determine the statistical significance of the results obtained, Shapiro–Wilks statistical test is applied to check the normality of the distributions and, as the results are not normally distributed, Mann–Whitney U rank test is used to asses if the pollutant is statistically reduced during Post-MC.

4. Pollution Evaluation at the LEZ Area

This section analyzes the information gathered by the sensor located in Pza. del Carmen (id. 35), which is the one inside the LEZ. When evaluating the pollution in MC, we face two main drawbacks: fFrst, the sensor does not gather information about $PM_{2.5}$ and PM_{10}, and second the noise data shared through ODP does not provide complete data for 2013. Thus, we do not include in this section the analysis of these two air pollutants and the Pre-MC period is defined from December 2014 to November 2018 when the noise is evaluated.

4.1. Air Quality

Table 2 reports the minimum (Min), the maximum (Max), the median (median), and the interquartile range (Iqr) for the concentration of the pollutants sensed in MC during the two periods analyzed here. The last column includes the value of Δ, and the check-mark (✓) indicates that there is a statistical difference according to the Mann–Whitney U test in such a pollutant for the periods analyzed (i.e., p-value < 0.01). Table 3 presents the percentage of time τ that the air pollutant concentration is lower than the threshold defined by the EU.

The evaluated measures are grouped by season because the meteorological conditions (i.e., wind direction and speed, atmospheric pressure, temperature, and relative humidity) affect the chemical behavior of the evaluated pollutants [65].

Figures 2 and 3 show the mean concentration of SO_2 and NO_2 by month, respectively. Notice that Pre-MC covers a longer amount of time. These figures also illustrate the boxplot of the concentration of the air pollutants for Pre-MC and Post-MC periods and the probability density function (PDF) of the whole data grouped by periods, i.e., Pre-MC and Post-MC. Figure 4 illustrates the regression analysis to evaluate the general trend of NO_2 through the time. Finally, Figures 5 and 6 show also the mean concentration of O_3 and CO by month, respectively.

Table 2. Summary of the air pollutants sensed. Negative values of Δ indicate a reduction of pollution and check-mark (✓) illustrates that there is statistical difference (p-value < 0.01).

Season	Pre-MC				Post-MC				Δ
	Min	Median	Iqr	Max	Min	Median	Iqr	Max	
SO_2 concentration in μg/m³									
spring	1.00	7.33	4.39	37.00	6.76	12.11	1.42	24.00	39.45
summer	1.00	8.90	5.94	50.00	1.00	10.49	1.98	17.00	15.19
autumn	1.00	8.33	5.53	53.00	1.00	11.69	6.38	35.00	28.70
winter	1.00	7.92	5.08	53.00	6.34	15.25	4.70	68.00	48.06
NO_2 concentration in μg/m³									
spring	5.00	38.39	20.30	162.00	1.00	23.92	16.39	131.00	✓ −60.47
summer	3.00	39.79	22.74	196.00	8.00	34.33	19.46	139.00	✓ −15.88
autumn	1.00	56.37	26.31	224.00	5.00	41.68	23.93	123.00	✓ −35.25
winter	4.00	51.05	25.43	196.00	1.00	49.63	26.00	147.00	−2.86
O_3 concentration in μg/m³									
spring	1.00	55.05	27.35	157.00	4.00	59.94	21.91	131.00	8.16
summer	1.00	63.40	32.83	215.00	1.00	52.74	26.45	135.00	✓ −20.20
autumn	1.00	26.49	22.80	134.00	1.00	31.77	21.26	112.00	16.61
winter	1.00	26.99	19.97	102.00	2.00	31.47	24.06	98.00	14.24
CO concentration in mg/m³									
spring	0.10	0.31	0.13	1.90	0.10	0.54	0.24	3.00	43.06
summer	0.10	0.33	0.21	2.30	0.10	0.27	0.28	3.00	✓ −21.38
autumn	0.10	0.47	0.30	2.60	0.10	0.50	0.26	2.00	5.31
winter	0.10	0.45	0.27	2.90	0.10	0.45	0.34	4.10	✓ 0.17

Table 3. Percentage of time τ the air pollutant concentration is lower than the EU thresholds.

Period	SO_2	NO_2	O_3	CO
Pre-MC	95.81	46.88	97.96	100.00
Post-MC	94.48	62.17	99.63	100.00

The concentration of SO_2 in the air is higher during Post-MC than during Pre-MC for all seasons (i.e., $\Delta > 0$). Figure 2 illustrates this increase. This is principally due to the largest sources of SO_2 emissions being fossil fuel combustion at power plants and other industrial facilities [65]. Thus, mobility-related measures or policies are not appropriate for reducing SO_2 concentration in the air.

For both periods, the mean concentration does not exceed the threshold defined by the EU (20 μg/m^3); however, the maxima values do (see Table 2). Table 3 shows that the population is under an SO_2 concentration lower than the UE threshold during 95.81% of the time for Pre-MC and 94.48% for Post-MC. Thus, the excess of this pollutant is exceptional and negligible, so it is not considered problematic in Spain.

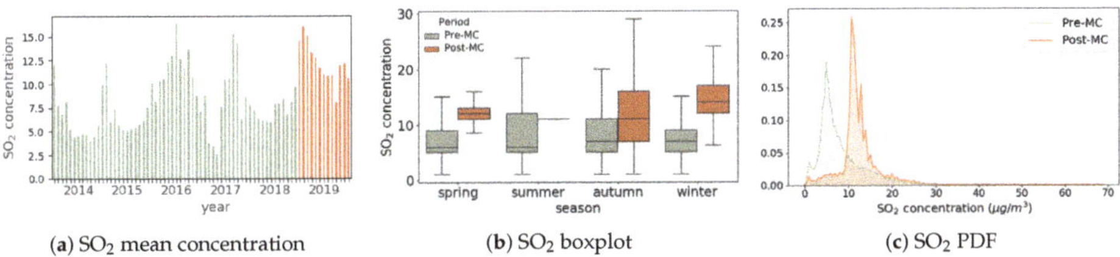

(**a**) SO_2 mean concentration (**b**) SO_2 boxplot (**c**) SO_2 PDF

Figure 2. Madrid Central SO_2: (**a**) Monthly mean concentration, (**b**) whole data concentration boxplot grouped by season, and (**c**) whole data PDF.

Focusing on NO_2, which is the pollutant that almost led Spain to the European Court and the excess of which is a public health concern, its concentration is significantly reduced for all seasons, except winter (i.e., Mann–Whitney U statistical test p-value < 0.01). The decrease of NO_2 concentration is lower during winter because of the heavier use of combustion power plants for wintertime home heating (therefore, road traffic may not be the main source of NO_2), and also because of the fact that NO_2 stays in the air longer in the winter [65]. Figure 3b confirms that warmer seasons have better air quality.

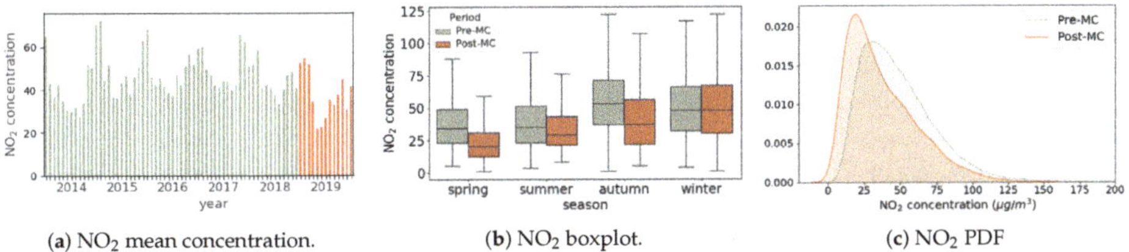

(**a**) NO_2 mean concentration. (**b**) NO_2 boxplot. (**c**) NO_2 PDF

Figure 3. Madrid Central NO_2: (**a**) Monthly mean concentration, (**b**) whole data concentration boxplot grouped by season, and (**c**) whole data PDF.

As can be seen in Figure 3a, the concentration of NO_2 exceeds, during several months, the maximum one allowed by the EU for both periods (Pre-MC and Post-MC) but with important differences. Thus, Table 3 results indicate that the population had healthier air for 15.29% longer duration during Post-MC as compared to Pre-MC circumstances (62.17%-46.88%). Figure 3c confirms that MC is under lower concentration of NO_2 in the air during Post-MC.

It is noticeable that there is a clear downward trend in the concentration of NO_2 after the application of the car restrictions (see Figure 3a). We apply regression analysis to evaluate the general trend of this pollutant. Figure 4 illustrates two different experiments: First, we take into account all the hourly concentration values during the last two years to see the trend of the closest time (a), and second the sensed values are averaged by month and the trends for the five years previous to the MC measurements (from 2014 to 2018) and also for the last five years (from 2015 to 2019) are evaluated (b). This last experiment analyzes the possible change of trend due to the new mobility policies.

In Figure 4a, the polynomial regression of grade 10 (black line) shows, first, how NO_2 concentration increases during colder seasons and decreases in warmer ones and, second, that the pollution values during Post-MC are lower than Pre-MC. In turn, the linear regression (black dashed line) displays a declined trend over time for this air pollutant. In Figure 4b, the black dashed line that represents the linear regression of NO_2 concentration before applying MC (five years) has a positive slope (i.e., the NO_2 concentration tends to increase). The solid black line that represents the general trend after applying MC measures has a negative slope, which indicates that NO_2 concentration in the air tends to be reduced. Thus, the behavior of the concentration of NO_2 in the air under the application of MC measures points out that the traffic restriction has a positive effect on air quality.

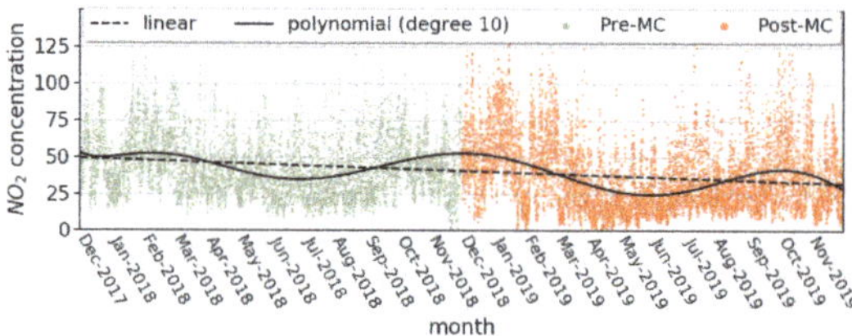

(**a**) Hourly values sensed during the last two years (from Dec 1st, 2017 to Nov 30th, 2019) grouped by hours.

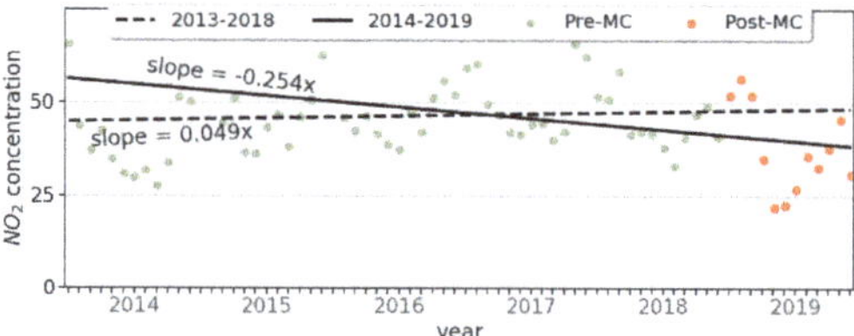

(**b**) Monthly values sensed during the last six years (from Dec 1st, 2013 to Nov 30th, 2019) grouped by month.

Figure 4. NO_2 values sensed at MC and the regression fitting.

The concentration of O_3 shows a similar behavior for both periods (see Figure 5 and Table 3). The results in Table 2 show that the concentration of this pollutant increased after the application of MC during spring, autumn, and winter, but it decreased during summer. All the monthly average O_3 values are lower than the maximum defined by the EU (120 μg/m^3). According to Table 3 values, the population is almost all the time experiencing air without an excess of O_3 during Post-MC.

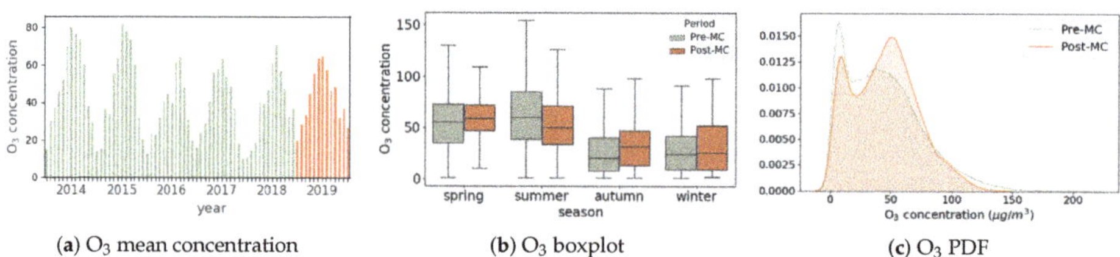

(**a**) O_3 mean concentration (**b**) O_3 boxplot (**c**) O_3 PDF

Figure 5. Madrid Central O_3: Mean concentration grouped by month (left side) and whole data concentration boxplot grouped by season (right side).

The increment of O_3 that we illustrate in our analysis could be due to the oxidation of NO, i.e., the chemical reaction of O_3 and NO that forms NO_2 and O_2, which occurs in urban areas [66]. As the road traffic limitation reduces the concentration of NO, the portion of O_3 that reacts with NO is lower. Therefore, the levels of O_3 do not decrease, and subsequently the concentration of NO_2 produced by the oxidation of NO is lower. In short, this upturn can be a chemical consequence of the reduction in the air of other components' concentrations.

Regarding CO, as it occurs with O_3, the concentration of this pollutant decreases during summer, but increases during the other seasons (see Table 2 and Figure 6). While one of the major sources of this pollutant concerning outdoor air is road traffic vehicles or machinery that burn fossil fuels, it seems that the reduction of road traffic does not lead to a decrease of CO in this case. However, according to

EU regulations, there is not a need to reduce CO since during the time frame analyzed in this article there is not any measurement over the threshold stipulated by the EU (10 mg/m^3).

Finally, the evaluation of SO_2, NO_2, O_3, and CO indicates that the final environmental balance may not always coincide with what was intuitively expected. However, it is important to remark that MC measures are highly effective in reducing NO_2 concentration, which was one of the major motivations for making this move. Therefore the answer to RQ1 (is the deployment of MC effective at reducing the concentration of NO_2 and so improving the air quality in this area?) is yes. The restriction to the road traffic applied in MC has significantly reduced NO_2 concentration.

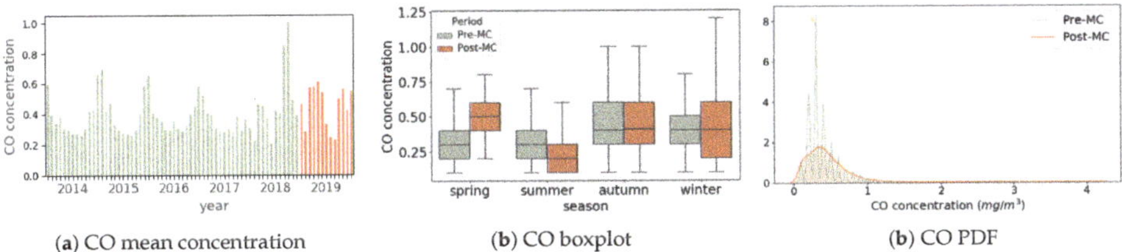

(a) CO mean concentration (b) CO boxplot (b) CO PDF

Figure 6. Madrid Central CO data: Mean concentration grouped by month (left side) and whole data concentration boxplot grouped by season (right side).

4.2. Outdoor Noise in MC Area

Table 4 reports the minimum (Min), the maximum (Max), the median (median), and the interquartile range (Iqr) of the levels of noise sensed in MC during the two periods analyzed here. The last column includes the value of Δ and the check-mark (✓) indicates that there is a statistical reduction in such a noise level between Pre-MC and Post-MC according to the Mann–Whitney U test (i.e., p-value < 0.01). Table 5 presents the percentage of time τ that the outdoor noise levels are lower than the threshold values defined by Madrid City Council [64] (see Section 3.2). Figures 7 and 8, and show the mean of the noise level evaluated here grouped by month.

Table 4. Summary of the levels of noise (in dBA). Negative values of Δ indicate a reduction of pollution and check-mark (✓) illustrates that there is statistical difference between the periods analyzed (p-value < 0.01).

Season	Pre-MC				Post-MC				Δ
	Min	Median	Iqr	Max	Min	Median	Iqr	Max	
L_{eq24}	59.30	67.50	1.70	95.80	62.30	66.70	1.62	77.30	✓ −0.34
L_{eqD}	58.80	68.20	2.20	84.00	60.70	67.50	2.22	80.10	✓ −0.22
L_{eqE}	57.10	67.80	1.67	82.00	57.80	67.20	1.50	82.60	✓ −0.46
L_{eqN}	55.10	64.90	2.50	80.70	58.50	64.40	2.33	78.40	✓ −0.55
L_{10}	62.70	70.70	1.70	81.50	66.40	70.00	1.60	86.70	✓ −0.37
L_{50}	55.00	65.30	2.00	71.60	57.50	64.40	2.10	71.50	✓ −0.71
L_{90}	47.30	56.90	2.50	66.30	49.80	56.20	2.50	63.40	✓ −0.77
NPL	68.70	80.90	3.00	108.40	74.90	80.10	2.80	100.50	✓ −0.00

Table 5. Percentage of time τ the levels of noise are lower than the EU thresholds.

Period	L_{eqE}	L_{eqD}	L_{eqN}
Pre-MC	4.94 (3.04)	8.30 (3.47)	0.00 (9.93)
Post-MC	3.02 (2.39)	9.89 (3.22)	0.00 (9.38)

Regarding the equivalent sound pressure levels (L_{eq24}, L_{eqD}, L_{eqE}, and L_{eqN}), the level the noise is higher during the hours between 7:00 h and 23:00 h than during the night time (see Table 4 and Figure 9). This is mainly due to the MC neighborhood being located in a commercial area and the business hours in Madrid usually ending at 22:00 h; thus, there is road traffic until late hours. There is a reduction in the median of all these noise levels during Post-MC. Therefore, in general, the noise is lower during this period. The highest differences between the two evaluated periods are given by the evening (L_{eqE}) and night (L_{eqN}) noise levels (see Table 4).

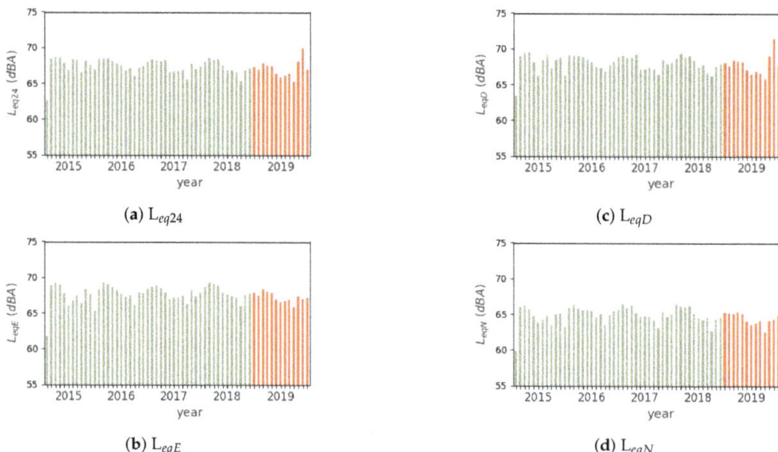

Figure 7. Mean levels of noise grouped by month.

Focusing on Post-MC time, the levels of noise decrease during the first months but they experience an increase after June, i.e., when MC suffered from the reversal (see Figure 7). If we focus on L_{eq24} and L_{eqD}, this increment is important. Taking into account the noise levels before the reversal, the reduction Δ for these two levels of noise is significantly higher, i.e., $\Delta(L_{eq24}) = -0.52$ dBA and $\Delta(L_{eqD}) = -0.45$ dBA. Thus, we can observe a negative impact on the noise pollution of the reversal that is mainly experienced at the time between 7:00h and 19:00h.

Evaluating τ during Pre-MC and Post-MC, Table 5 shows that the outdoor levels of noise practically always surpass the thresholds of the city council. During the night the levels of noise are never lower than 55 dBA, which is the threshold for these hours. Notice that we are evaluating equivalent sound pressure levels averaged for every day. Figure 7 also illustrates how the noise levels generally exceed the thresholds, i.e., the monthly average levels of noise are higher than 65 dBA for L_{eqD} and L_{eqE} and higher than 55 dBA for L_{eqN}. Therefore, other measures must be applied to further reduce outdoor noise in this area.

Focusing on the percentile noise levels (L_{10}, L_{50}, and L_{90}), Pre-MC and Post-MC differences are statistically significant. The best improvement Δ is shown by L_{90} (see Table 4), which represents the residual background levels of noise of the urban area analyzed. As the continuous road traffic flow is one of the main sources of the background noise, the reduction of traffic transit provokes a decrease in this type of noise.

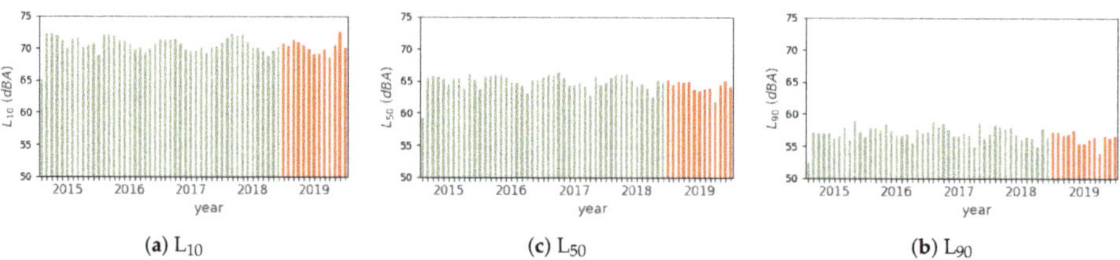

Figure 8. Mean percentile levels of noise grouped by month.

L_{50}, which statistically represents the median of the fluctuating levels of noise, is also reduced, i.e., $\Delta(L_{50}) = -0.71$. The reduction of the annoying peaks in noise (i.e., L_{10}) is lower than for the other two percentile levels ($\Delta(L_{10}) = -0.37$). This represents a limited decrease of 0.5% regarding the median value of this noise during Pre-MC (70.70 dBA).

There is not a statistically significant reduction of NPL (see Table 4) and the average improvement is 0.00. Besides, as the computation of this metric depends on L_{eq24}, it undergoes the same increase during the last months of Post-MC after the reversal of MC (see Figure 10).

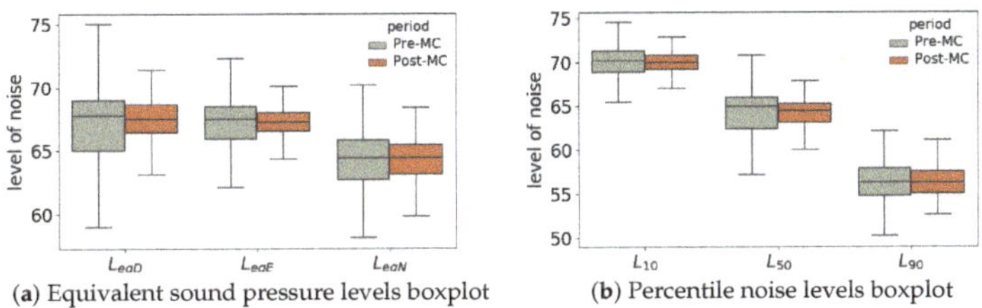

(**a**) Equivalent sound pressure levels boxplot (**b**) Percentile noise levels boxplot

Figure 9. Boxplots of the levels of noise evaluated gruoped by periods.

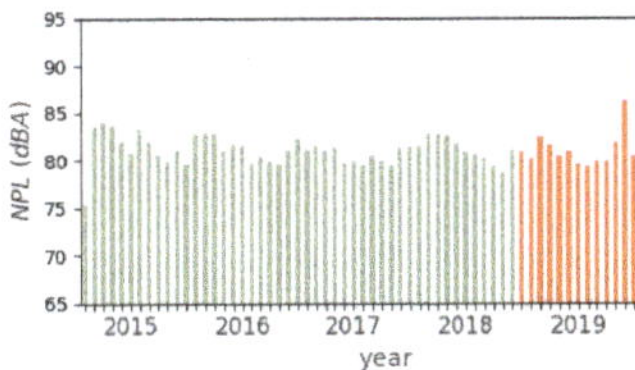

Figure 10. Madrid Central noise pollution level (NPL) monthly levels of noise.

Figures 11–13 illustrate the trend of some representative levels of sound (L_{eq24}, L_{eqD}, and L_{90}). When evaluating the trend of the last two years (on top of each figure), we observe similar behavior: (a) The polynomial regression of grade 10 (black line) illustrates the most marked reduction of the levels of noise during Post-MC until June 2019 when the MC was reversed, and (b) the linear regression (black dashed line) shows that the noise in MC is marginally reduced over time (i.e., the function presents negative slopes but close to zero).

At the bottom of Figures 11–13, the black dashed lines that represents the linear regression of the levels of noise before applying MC (from 2014 to 2018) have slopes close to zero or even positive in the case of L_{eqD}, i.e., there is an increase of outdoor noise in the area. The solid black line that represents the general trend after applying MC measures (last four years) has a steeper negative slope, which is higher in the case of L_{90}. Thus, the car restrictions tend to improve the background noise generated by road traffic.

Finally, we have computed the PDF of the L_{eq24}, L_{eqD}, and L_{90} noise values to confirm that there is a slight reduction in the outdoor levels of noise in the area of MC. Figure 14 illustrates that the distributions of values detected Post-MC are more likely to be lower than the Pre-MC ones (the Post-MC distribution is lightly shifted to the left). However, while there is such a reduction, it is notorious that other types of measures are needed to mitigate this source of health problems and discomfort because the levels of noise exceed the thresholds set by the institutions nearly all the time during both evaluated periods, Pre-MC and Post-MC.

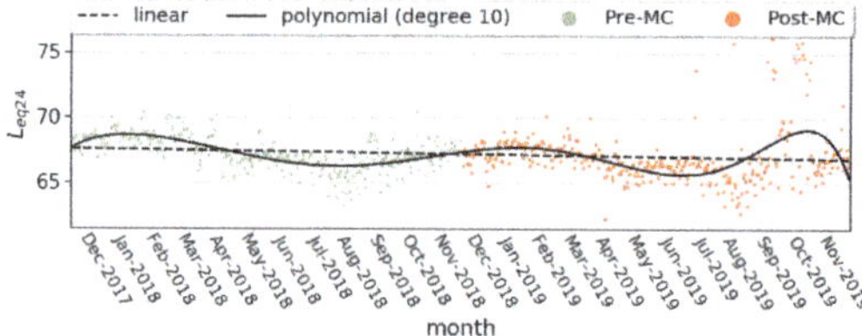

(**a**) Values detected during the last two years (from 1 December 2017 to 30 November 2019) grouped by hours.

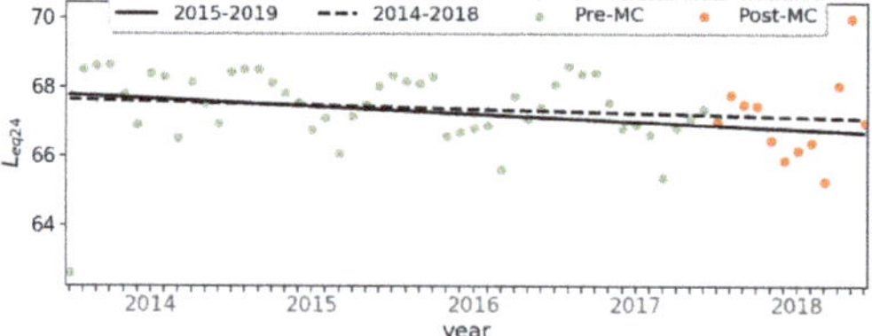

(**b**) Values detected during the last five years (from 1 December 2014 to 30 November 2019) grouped by month.

Figure 11. Madrid Central sensor data. L_{eq24} values sensed and regression fitting.

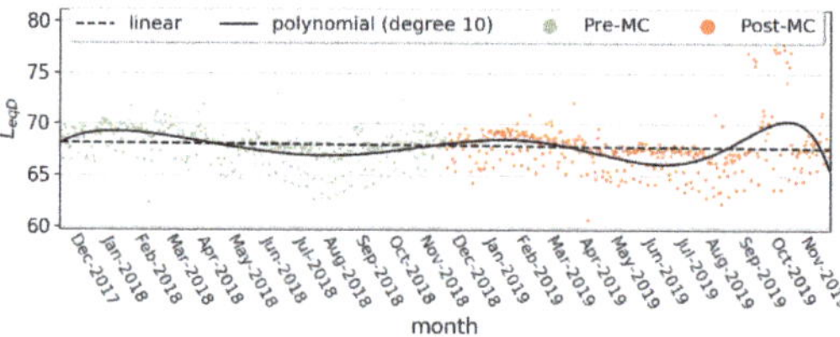

(**a**) Values sensed during the last two years (from 1 December 2017 to 30 November 2019) grouped by hours.

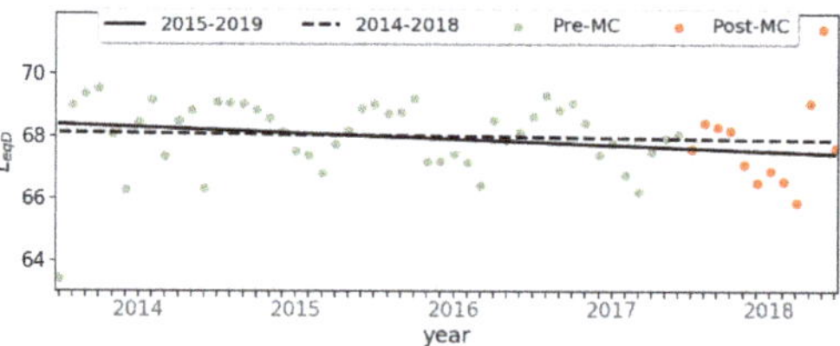

(**b**) Values sensed during the last five years (from 1 December 2014 to 30 November 2019) grouped by month.

Figure 12. Madrid Central sensor data. L_{eqD} values sensed and regression fitting.

According to these results, we answer RQ2: Is the definition of an LEZ an effective measure to reduce environmental noise levels?. MC slightly reduces the outdoor levels of noise, mainly the background noise produced by road traffic. However, this decrease is not enough to keep the noise in the range of healthy levels.

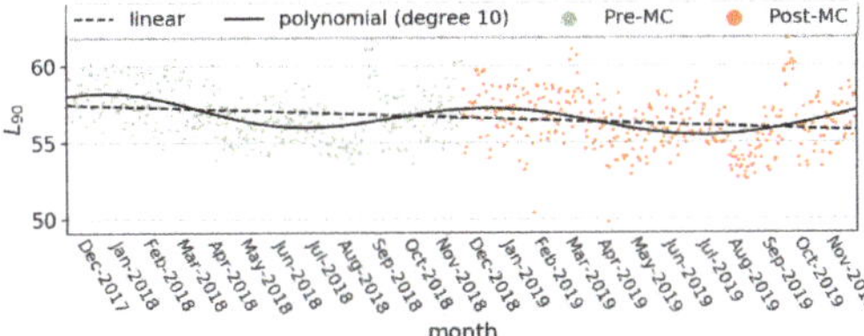

(**a**) Values sensed during the last two years (from 1 December 2017 to 30 November 2019) grouped by hours.

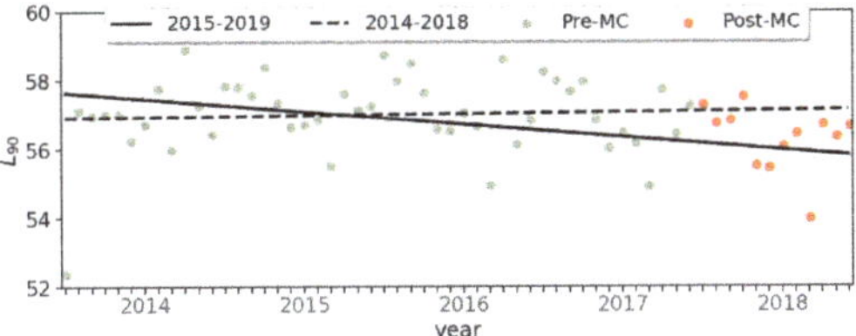

(**b**) Values sensed during the last five years (from 1 December 2014 to 30 November 2019) grouped by month.

Figure 13. Madrid Central sensor data. L_{90} values sensed and regression fitting.

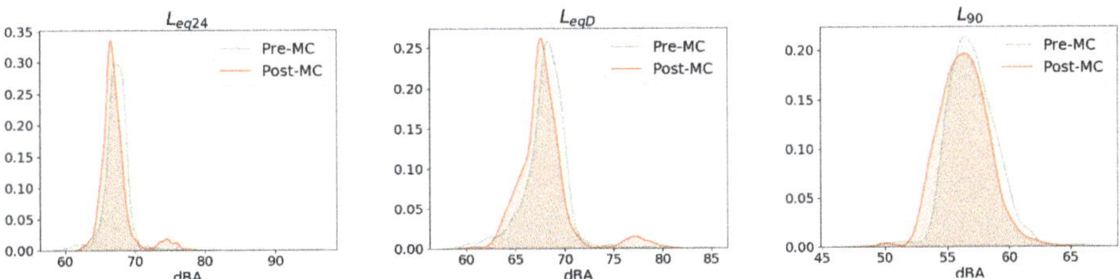

Figure 14. Madrid Central noise. PDF of the L_{eq24}, L_{eqD}, and L_{90} noise levels.

5. Indirect Repercussion on the Pollution of the Whole City

According to the WHO and the EU, NO_2 and particulate matter (PM_2.5 and $PM_1$0) are the main culprits in public health problems due to pollution. This section aims at taking advantage of the sensors network installed in Madrid to:

- First, confirm that the important reduction of NO_2 emissions in the MC area does not lead to an increase of that pollutant in other zones due to a hypothetical redirection of traffic to other areas of the city, i.e., to investigate the possible border effect of MC, and
- second, analyze the concentration of the particulate matter in the areas where the sensors gather such information in order to asses the possible impact on this aspect of the city due to MC measures.

The data provided by the ODP is incomplete and contains errors for several of the areas covered by the sensors. Only the data coming from sensors that correctly registered values for the whole time frame analyzed here are used in this analysis. Therefore, this section discusses data about NO_2 from 23 sensors, $PM_{2.5}$ from six, and PM_{10} from 12. This limits the areas of Madrid analyzed in our article. However, it ensures that the data reflect reliably the concentration in the air of the pollutants analyzed.

5.1. Temporal Variation in the Air Pollutants

In order to better understand the impact of NO_2, PM_2.5, and $PM_1$0 in Madrid, we evaluate their monthly trends that are shown in Figure 15. As seen in Section 4.1, NO_2 exhibits seasonal variations, with the highest concentrations in winter and the lowest in summer. During warmer months (from March to September) the median NO_2 concentration is lower than the EU threshold (40 $\mu g/m^3$). The highest NO_2 concentration occurs during December and the lowest one during August.

This seasonal variation is mainly due to two different factors: The meteorological conditions and the emissions patterns. For example, temperature inversions and lower boundary layer heights in winter can cause NO_2 not to be ventilated from the boundary layer, leading to higher concentrations in European cities. In contrast, the increase of photochemical activity, solar radiation, etc., during summer lowers NO_2 concentration [67]. In addition, fossil fuel combustion sources such as residential coal and biomass combustion for heating also contributed to the formation of high NO_2 concentration in wintertime.

Focusing on $PM_{2.5}$, the highest concentration of this pollutant occurs in December (as with NO_2). This pollutant presents a decreasing trend from December to May, springtime being the least polluted season. March, April, and May median values are the only ones that are lower than the threshold marked by the EU. For the months between June and November, the $PM_{2.5}$ concentrations tend to be slightly higher than the EU threshold and similar to each other. PM_{10} shows a similar seasonal variation to $PM_{2.5}$; from December to May there is a decrease in the concentration of this pollutant. July is the month with the highest PM_{10} concentration and May the month with the lowest one. In the case of this pollutant, June, July, August, and October median values surpass the EU threshold (40 $\mu g/m^3$).

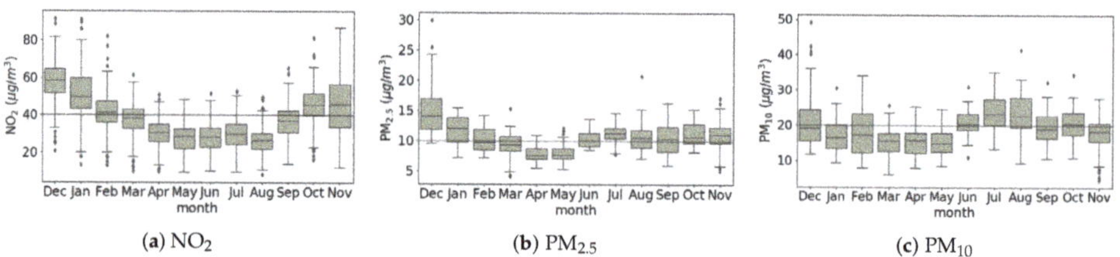

(**a**) NO_2 (**b**) $PM_{2.5}$ (**c**) PM_{10}

Figure 15. Monthly variations of NO_2, $PM_{2.5}$, and PM_{10}. The black horizontal lines represent the limits specified by the EU.

5.2. *Effect of Mobility Restrictions on* NO_2 *Concentration in Other Areas of the City*

The application of restrictions to vehicles in accessing a street or area affects the mobility patterns of the inhabitants in the whole city [3], and therefore impacts on the pollution in different areas of that city. One of the main motivations of the deployment of MC was the reduction of this pollutant (see Section 2.2). In order to asses the effect of MC on the rest of the city of Madrid, we study NO_2 concentration measured by sensors located through Madrid that gather the ODP data (see Figure 1). Thus, we evaluate the concentration of this pollutant during Pre-MC and Post-MC. Thus, we compute the difference between these two periods for each one of the sensed areas.

Table 6 shows the median (Med) and interquartile range (Iqr) of the daily sensed NO_2 concentration for each sensor grouped by year (these values are selected because the distributions are not normally distributed). The last column presents the Δ values. The minimum median concentration for each sensor is marked in bold. Finally, Figure 16a illustrates the boxplots of the concentration for all the sensors for each year.

According to the results in Table 6, the median NO_2 concentration in the air during the year 2019 (i.e., Post-MC period) is the minimum for 14 of the 23 areas sensed. It is noticeable that 2014 is the year that recorded the minimum median concentration of this pollutant (for 11 of the 23 sensors). Thus, the years with lowest NO_2 concentration are first 2019 and second 2014 (see Figure 16a).

During the period 2015 to 2018, several areas suffered from median NO_2 concentrations higher than 40 $\mu g/m^3$ (the EU threshold), which motivated the EU to fine Spain and to apply the subsequent measures to define MC to avoid it. In 2019 and 2014 only the Pza. Elíptica sensor exceeded this threshold. The main problem in this area is that it suffers from heavy traffic and congestion because its main road, the A-41 highway, connects Madrid with the southern towns.

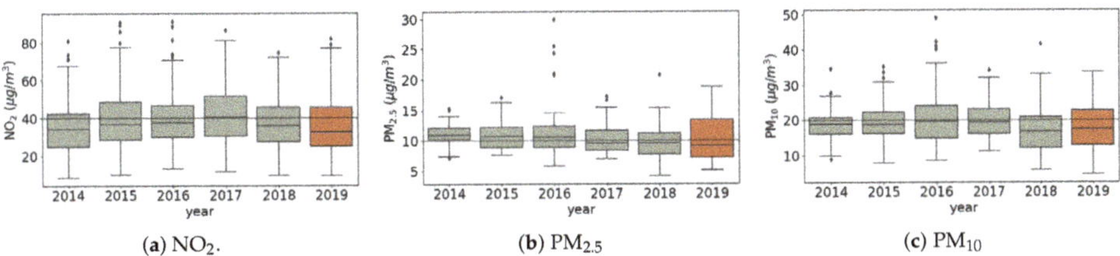

(**a**) NO_2. (**b**) $PM_{2.5}$ (**c**) PM_{10}

Figure 16. The city-wide daily concentrations of the evaluated pollutants during the study period grouped by year. The black horizontal lines represent the limits specified by the EU.

Table 6. Median NO_2 concentration and Δ values for the sensors analyzed. The minimum median concentration for each sensor is marked in bold.

Sensor Location	**Years**												**Δ**
	2014		**2015**		**2016**		**2017**		**2018**		**2019**		
	Med	**Iqr**	**Med**	**Iqr**	**Med**	**Iqr**	**Med**	**Iqr**	**Med**	**Iqr**	**Med**	**Iqr**	
Pza. del Carmen	34.0	34.0	44.0	37.0	44.0	31.0	44.0	32.0	41.0	28.0	**31.0**	32.0	✓ −8.87
Pza. de España	**34.0**	31.0	44.7	34.0	41.0	37.0	43.0	39.0	38.0	34.0	35.0	35.0	✓ −5.48
Avd. Ramón y C.	31.0	38.0	38.0	39.0	38.5	37.0	36.0	38.6	35.0	38.0	**33.0**	37.0	✓ −2.50
Arturo Soria	29.0	33.0	33.0	40.0	32.0	34.0	32.0	38.0	28.0	33.0	**27.0**	33.2	✓ −3.51
Villaverde	**27.0**	36.0	31.0	46.0	33.0	41.0	35.0	49.0	**27.0**	38.0	**27.0**	44.0	✓ −2.12
Farolillo	26.0	32.0	31.0	39.0	32.0	39.0	32.0	40.0	26.0	33.0	**25.0**	35.0	✓ −3.86
Casa de Campo	**12.0**	24.0	15.0	28.0	15.0	26.0	16.3	28.0	**12.0**	24.0	**12.0**	26.0	✓ −0.94
Barajas Pueblo	**25.0**	33.0	26.0	36.0	31.0	39.0	31.0	42.0	29.0	36.0	28.0	38.0	1.04
Moratalaz	**28.0**	30.0	33.0	36.0	34.0	33.0	33.0	36.0	30.5	34.0	29.0	36.0	✓ −1.48
Cuatro Caminos	35.0	35.0	36.0	41.0	35.0	37.0	37.0	40.0	34.0	37.0	**31.0**	38.0	✓ −3.13
Barrio del Pilar	31.0	36.0	32.0	40.1	33.0	35.0	32.0	39.0	30.0	35.0	**26.5**	38.0	✓ −3.79
Vallecas	33.0	35.0	29.0	37.0	32.0	38.0	32.0	39.0	28.0	33.0	**28.0**	36.0	✓ −2.93
Mendez Alvaro	**24.0**	35.0	29.0	41.0	30.0	41.0	33.0	45.0	26.0	36.0	26.0	37.0	✓ −2.89
Castellana	34.0	34.0	33.0	36.0	35.0	34.0	32.0	37.0	32.0	35.0	**28.0**	36.0	✓ −3.30
Par. del Retiro	19.0	22.0	25.0	31.0	26.0	33.0	23.2	35.0	22.0	27.0	**18.0**	29.0	✓ −4.34
Plaza Castilla	39.0	38.0	40.0	43.0	40.0	36.0	34.0	36.0	33.0	36.0	**31.0**	35.0	✓ −4.76
Ens. de Vallecas	**21.0**	28.0	31.0	43.0	27.0	40.0	28.0	40.0	27.0	33.0	27.0	37.0	0.75
Urb. Embajada	**31.0**	39.0	37.0	44.0	37.0	40.2	39.0	47.0	33.0	41.0	**31.0**	43.0	✓ −2.89
Pza. Elíptica	**46.0**	42.0	49.0	41.0	49.0	41.0	48.0	45.0	46.0	42.0	47.0	42.0	✓ −0.97
Sanchinarro	24.0	28.0	24.0	33.0	27.0	30.0	23.0	32.0	**21.0**	29.0	23.0	30.0	✓ −0.86
El Pardo	**9.0**	15.0	13.0	18.0	14.0	17.0	13.0	19.0	10.0	16.0	12.0	16.0	−0.02
Juan Carlos I	**14.0**	22.0	16.0	26.0	16.0	23.0	18.0	28.0	17.0	26.0	18.0	28.0	3.74
Tres Olivos	21.0	29.0	27.0	38.0	27.0	36.0	25.0	37.0	20.0	29.0	**18.0**	22.0	✓ −8.45

As shown in the last column of Table 6, the highest reduction of NO_2 concentration was detected in Pza. del Carmen, which is in the MC area. As expected, another important reduction occurred in the area monitored by the sensor installed in Pza. España (sensor id 4), which is the closest sensor to the MC area. It experienced a reduction in NO_2 concentration of 5.48 $\mu g/m^3$. In addition, there was a reduction in the average NO_2 concentration for all the sensed areas but three exceptions (Barajas Pueblo, Ens. de Vallecas, and Juan Carlos I, which are suburb areas far from the center of Madrid).

The results in Table 6 and Figure 16a indicate that, in general, the deployment of the LEZ has a positive impact on the whole city because, after its implementation, the air in Madrid generally is healthier (contains less NO_2). These results are in line with the study that acknowledged that NO_2 concentration levels in Madrid are dominated by local traffic (up to 90%) [57]. Thus, reducing the road traffic leads to a reduction in NO_2 concentration in this city.

5.3. *Repercussion on the Particulate Matter Concentration in Other Areas of Madrid*

As the sensor located at MC does not gather particulate matter concentration data, we analyze the effect of MC on this type of pollutant in different areas of the city. The number of sensors that gather trustworthy data during the time frame of our study is only six for $PM_{2.5}$ and 12 for PM_{10}. Figure 17 shows the location of these sensors. This limits the outcomes of this section about the concentration of these pollutants in the whole city.

Tables 7 and 8 present the median (Med) and interquartile range (Iqr) of the daily sensed concentrations of $PM_{2.5}$ and PM_{10}, respectively, for each sensor grouped by year. The last column presents the Δ values (difference between Pre-MC and Post-MC). The minimum median concentration for each sensor is marked in bold. Finally, Figure 16b,c show the boxplots of the concentration for all the sensors for each year for $PM_{2.5}$ and PM_{10}, respectively.

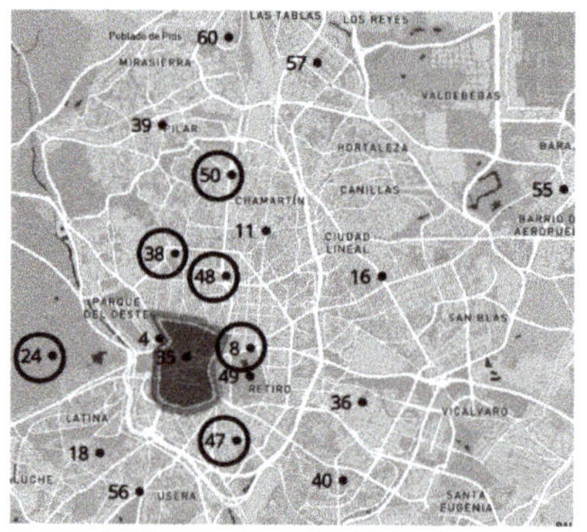

(a) Sensors that gather $PM_{2.5}$ data

(a) Sensors that gather PM_{10} data

Figure 17. Location of the sensors that gather particulate matter concentration data.

Table 7. Median $PM_{2.5}$ concentration and Δ values for the sensors analyzed. The minimum median concentration for each sensor is marked in bold.

Sensor Location	Years												Δ
	2014		2015		2016		2017		2018		2019		
	b	Iqr	Med	Iqr	Med	Iqr	Med	Iqr	Med	Iqr	Med	Iqr	
Casa de Campo	**7.0**	6.0	8.0	9.0	8.0	9.0	9.0	9.0	6.0	8.0	8.0	8.0	−0.06
Cuatro Caminos	10.0	11.0	9.6	8.0	9.0	9.0	9.0	8.0	**8.0**	8.0	**8.0**	8.0	✓ −0.49
Mendez Alvaro	**9.0**	8.0	**9.0**	9.0	**9.0**	11.0	10.0	10.0	**9.0**	10.0	**9.0**	9.0	✓ −0.68
Castellana	**8.0**	9.0	9.0	4.0	9.0	9.0	8.0	7.0	8.0	8.0	**8.0**	8.0	✓ −0.05
Plaza Castilla	10.0	7.0	9.0	8.0	9.0	9.0	8.0	7.0	8.0	8.0	**7.0**	8.0	✓ −0.75
Escuelas Aguirre	10.0	7.6	11.0	9.0	10.0	10.0	10.0	8.6	**9.0**	9.0	**9.0**	10.0	✓ −0.80

There are only six sensors able to trustworthily provide data about $PM_{2.5}$ during the time frame of our study and they are located in the downtown of the city (see Figure 17a). Five of these six sensed areas show the minimum concentration of of this pollutant during 2019 (see Table 7). The sensor that does not presentsits minimum during 2019 (Casa de Campo sensor id 24) is located in the center of a large park (green area). In addition, Casa de Campo presents the lowest $PM_{2.5}$ concentration in comparison with the others. Among the analyzed years, the median of the $PM_{2.5}$ concentration during 2019 is the lowest (see Figure 16b). For all the evaluated areas, there is a decrease in the concentration of this pollutant, which is statistically significant for five of them. This is expected because the main source of $PM_{2.5}$ in Madrid is the road traffic according to the Screening for High Emission Reduction Potentials for Air quality tool (SHERPA) developed by the Joint Research Centre to quantify the origins of air pollution in cities and regions [68].

Table 8. Median PM_{10} concentration and Δ values for the sensors analyzed. The minimum median concentration for each sensor is marked in bold.

Sensor Location	Years												Δ
	2014		2015		2016		2017		2018		2019		
	Med	Iqr	Med	Iqr	Med	Iqr	Med	Iqr	Med	Iqr	Med	Iqr	
Farolillo	15.0	15.0	18.0	16.0	15.0	16.0	22.0	20.0	14.0	17.0	**13.9**	13.0	✓ −4.12
Casa de Campo	13.0	12.0	15.5	14.0	14.0	16.0	17.0	16.0	12.0	14.1	**11.0**	12.0	✓ −4.19
Moratalaz	**17.0**	21.0	18.0	21.0	**17.0**	20.0	20.0	17.0	18.0	17.0	**17.0**	16.7	−1.00
Cuatro Caminos	**14.0**	12.0	16.0	13.0	16.0	17.0	16.0	15.0	**14.0**	16.7	17.0	16.0	1.10
Vallecas	14.0	14.0	15.0	18.0	13.0	17.0	15.0	19.0	14.0	17.0	**13.0**	16.0	✓ −1.82
Mendez Alvaro	15.0	14.0	16.0	15.0	16.0	17.0	16.0	14.0	**13.0**	15.0	16.0	15.5	−0.48
Castellana	16.0	19.0	15.0	13.0	15.0	16.0	14.0	13.0	**12.0**	15.0	15.0	15.0	0.11
Plaza Castilla	16.0	14.0	14.0	15.0	13.0	17.0	11.0	13.0	**13.0**	14.0	16.0	15.0	2.24
Urb. Embajada	**14.0**	17.0	16.0	21.0	16.0	21.0	19.0	20.0	17.0	18.0	19.0	20.0	1.27
Sanchinarro	13.0	16.0	14.0	20.0	14.0	20.0	15.0	21.0	12.0	16.0	**12.0**	13.0	✓ −4.48
Tres Olivos	11.0	17.0	14.0	21.0	13.0	19.0	13.0	22.0	13.0	18.0	**10.0**	14.0	✓ −4.76
Escuelas Aguirre	19.0	16.0	20.0	17.0	18.0	19.0	18.0	15.0	**15.0**	18.0	19.0	18.0	−0.62

The PM_{10} concentration in the air for the 12 sensed areas does not show a clear trend. Six of these 12 areas show the minimum concentration of of this pollutant during 2019 and five during 2018 (see Table 8). The concentration of PM_{10} during these two years (2018 and 2019) shows lower distributions similar to the other periods of time (see Figure 16c). When comparing Pre-MC and Post-MC, eight areas present a decrease in PM_{10} after the deployment of MC, five of which are statistically lower.

According to the results about NO_2, $PM_{2.5}$, and PM_{10} concentrations in the whole city of Madrid, the answer to RQ3 (do pedestrianization policies in a given area of the city produce a pollution displacement to other zones of the city?) is that they do not produce pollution displacement. These pedestrianization policies positively impact on the whole city because there is a general reduction in the concentration of these three pollutants.

Finally, after the whole study we answer RQ4: Are smart city tools effective for evaluating urban health policies and other measures implemented in the city?. As we have seen throughout the article, the application of smart city tools proves to be an effective way of assessing the effectiveness of measures against urban pollution. In this sense, the evolution towards the smart city would improve the local capacity to deal with the risks arising from rapid urbanization. The application of Internet of Things to policy implementation allows us to avoid wrong, subjective, or biased appraisal, offering an objective assessment of their effectiveness. Furthermore, the smart city paradigm proves to be the best way to monitor compliance with international emission requirements. However, we have faced the issue of dealing with non-homogeneous and incomplete data for our study. This may limit the outcomes of the analysis carried out: Data analyses only can be as trustworthy as the data source. Thus, it is mandatory to provide a platform able to gather and share complete and accurate data.

6. Conclusions

The growth of car-oriented cities is raising new urban health problems resulting from the pollution increase. This requires quick responses to create sustainable environments from an environmental point of view. Several initiatives are being taken into account to address this challenge but some have been questioned in terms of their effectiveness. Smart city related technologies provide invaluable tools of analysis, helping decision making, and leading to the best outcome for the city. In this article, we evaluate the LEZ deployed in Madrid (Spain), applying smart city tools in order to objectively asses the reduction of the pollution of this measure and the potential side-effects.

Real data provided by the Madrid City Council was processed to get time series of air pollutant concentrations and levels of noise in different areas of the city. According to the statistical and regression analyses, MC was able to significantly reduce NO_2 concentration locally, having the same positive impact on the rest of the city. In addition, a decrease in $PM_{2.5}$ and PM_{10} has occurred in most of the analyzed zones of the city. Thus, this LEZ effectively improves the air quality and does not provoke the border effect. In terms of noise, this measure is able to slightly reduce outdoor noise levels, mainly the background ones generated by road traffic.

We found difficulties in terms of the quantity, quality, and reliability of the open data shared by the city council. Despite these limitations, smart city tools in Madrid have proved to be an invaluable resource to evaluate the effectiveness of this type of environmental measure.

The main lines for future work include extending the analysis, performing a multivariate analysis by taking into account related data (e.g., wind speed, temperature, etc.); evaluating the impact on other relevant indicators (e.g., economical impact, mobility behavior, citizens' health, etc.); and applying other time series analysis methods and models (e.g., Markov Chains and recurrent neural networks) to characterize the pollution.

Author Contributions: Conceptualization, J.T. and I.L.; methodology, J.T. and I.L.; software, J.T.; formal analysis, J.T.; investigation, J.T. and I.L.; resources, J.T. and I.L.; data curation, J.T.; writing–original draft preparation, J.T. and I.L.; writing–review and editing, J.T. and I.L; visualization, J.T.; supervision, J.T.; project administration, J.T. and I.L.; funding acquisition, J.T. and I.L. All authors have read and agreed to the published version of the manuscript.

Funding: Irene Lebrusán was partially funded by RCC Postdoctoral Research Fellowships at Harvard University for Distinguished Junior Scholars. Jamal Toutouh has been partially funded by European Union's Horizon 2020 research and innovation programme under the Marie Skłodowska-Curie grant agreement No 799078. This research has been partially funded by the Spanish MINECO and FEDER projects TIN2017-88213-R (http://6city.lcc.uma.es) and TIN2016-81766-REDT (http://cirti.es). Universidad de Málaga, Campus Internacional de Excelencia Andalucía TECH.

Acknowledgments: Irene Lebrusán was partially funded by RCC Postdoctoral Research Fellowships at Harvard University for Distinguished Junior Scholars. Jamal Toutouh has been partially funded by European Union's Horizon 2020 research and innovation programme under the Marie Skłodowska-Curie grant agreement No. 799078. This research has been partially funded by the Spanish MINECO and FEDER projects TIN2017-88213-R (http://6city.lcc.uma.es) and TIN2016-81766-REDT (http://cirti.es). Universidad de Málaga, Campus Internacional de Excelencia Andalucía TECH. Consejería de Economía y Conocimiento de la Junta de Andaluía, UMA18-FEDERJA-003 (Precog).

Conflicts of Interest: The authors declare no conflict of interest.

References

1. United Nations. World Urbanization Prospects: The 2018 Revision: Key Facts. 2018. Available online: https://population.un.org/wup/Publications/Files/WUP2018-KeyFacts.pdf (accessed on 7 July 2019).
2. World Health Organization *Promoting Physical Activity and Active Living in Urban Environments*; World Health Organization: Geneva, Switzerland, 2006.
3. Soni, N.; Soni, N. Benefits of pedestrianization and warrants to pedestrianize an area. *Land Use Policy* **2016**, *57*, 139–150. [CrossRef]
4. Basner, M.; Babisch, W.; Davis, A.; Brink, M.; Clark, C.; Janssen, S.; Stansfeld, S. Auditory and non-auditory effects of noise on health. *Lancet* **2014**, *383*, 1325–1332. [CrossRef]
5. City, B.L.; Assessment, E. Urbanization and health. *Bull. World Health Organ.* **2010**, *88*, 245–246.
6. Steele, C. A critical review of some traffic noise prediction models. *Appl. Acoust.* **2001**, *62*, 271–287. [CrossRef]
7. Parajuli, A.; Pojani, D. Barriers to the pedestrianization of city centres: Perspectives from the Global North and the Global South. *J. Urban Des.* **2018**, *23*, 142–160. [CrossRef]
8. Ward, S.V. What did the Germans ever do for us? A century of British learning about and imagining modern town planning. *Plan. Perspect.* **2010**, *25*, 117–140. [CrossRef]
9. Tobon, M.; Jaramillo, J.P.; Sarmiento, I. Pedestrianization and semi-pedestrianization: A model for recovery public space in the Medellín downtown. In Proceedings of the MOVICI-MOYCOT 2018: Joint Conference for Urban Mobility in the Smart City, Medellin, Colombia, 18–20 April 2018; pp. 1–7. [CrossRef]
10. Sobková, L.F.; Čertický, M. Urban Mobility and Influence Factors: A Case Study of Prague. *WIT Trans. Built Environ.* **2017**, *176*, 207–217.

11. EU regional and Urban Development, European Commission. Smart Cities. 2018. Available online: https://ec.europa.eu/info/eu-regional-and-urban-development/topics/cities-and-urban-development/city-initiatives/smart-cities_en (accessed on 5 January 2020).
12. Lebrusán, I.; Toutouh, J. *Assessing the Environmental Impact of Car Restrictions Policies: Madrid Central Case*; Ibero-American Congress on Information Management and Big Data; Springer: Berlin, Germany, 2019; pp. 9–24.
13. Lynch, K.; Banerjee, T. *Growing up in cities: Studies of the spatial environment of Adolescence in Cracow, Melbourne, Mexico City, Salta, Toluca, and Warszawa*; MIT Press: Cambridge, MA, USA, 1977.
14. Tonucci, F. Citizen child: Play as welfare parameter for urban life. *Topoi* **2005**, *24*, 183–195. [CrossRef]
15. Fotel, T.; Thomsen, T.U. The Surveillance of Children's Mobility. *Surveill. Soc.* **2003**, *1*. [CrossRef]
16. Lawton, M.P.; Nahemow, L. *Ecology and the Aging Process*; American Psychological Association: Washington, DC, USA, 1973.
17. World Health Organization. Ambient (Outdoor) Air Quality and Health. 2018. Available online: https://www.who.int/en/news-room/fact-sheets/detail/ambient-(outdoor)-air-quality-and-health (accessed on 7 July 2019).
18. European Environment Agency. Air Quality in Europe—2018. 2018. Available online: https://www.eea.europa.eu/publications/air-quality-in-europe-2018 (accessed on 7 July 2019).
19. Raaschou-Nielsen, O.; Andersen, Z.J.; Beelen, R.; Samoli, E.; Stafoggia, M.; Weinmayr, G.; Hoffmann, B.; Fischer, P.; Nieuwenhuijsen, M.J.; Brunekreef, B.; et al. Air pollution and lung cancer incidence in 17 European cohorts: prospective analyses from the European Study of Cohorts for Air Pollution Effects (ESCAPE). *Lancet Oncol.* **2013**, *14*, 813–822. [CrossRef]
20. Brivio, E.; Meder, S. Environmental Implementation Review: Commission Helps Member States to Better Apply EU Environment Rules to Protect Citizens and Enhance Their Quality of Life. 2019. Available online: https://europa.eu/rapid/press-release_IP-19-1934_en.htm (accessed on 7 September 2019).
21. Younan, D.; Petkus, A.J.; Widaman, K.F.; Wang, X.; Casanova, R.; Espeland, M.A.; Gatz, M.; Henderson, V.W.; Manson, J.E.; Rapp, S.R.; et al. Particulate matter and episodic memory decline mediated by early neuroanatomic biomarkers of Alzheimer's disease. *Brain* **2020**, *143*, 289–302. [CrossRef]
22. Stansfeld, S.A.; Matheson, M.P. Noise pollution: Non-auditory effects on health. *Br. Med Bull.* **2003**, *68*, 243–257. [CrossRef]
23. Evans, G.W. Child development and the physical environment. *Annu. Rev. Psychol.* **2006**, *57*, 423–451. [CrossRef]
24. Luxon L. M.; Prasher D. *Noise and Its Effects*; Wiley: New York, NY, USA, 2007.
25. European Environment Agency. Noise Pollution Is a Major Environmental Health Concern in Europe. 2016. Available online: https://www.eea.europa.eu/themes/human/noise (accessed on 7 July 2019).
26. Goines, L.; Hagler, L.C.M. Noise Pollution: A Modern Plague. *South Med J.* **2007**, *100*, 287–294. [CrossRef]
27. Silva, L.T.; Mendes, J.F. City Noise-Air: An environmental quality index for cities. *Sustain. Cities Soc.* **2012**, *4*, 1–11. [CrossRef]
28. European Commission. Directive 2008/50/EC of the European Parliament and of the Council of 21 May 2008 on ambient air quality and cleaner air for Europe. *Off. J. Eur. Union L* **2008**, *152*, 1–44.
29. European Commission. Directive 2004/107/EC of the European Parliament and of the Council of 15 December 2004 relating to arsenic, cadmium, mercury, nickel and polycyclic aromatic hydrocarbons in ambient air. *Off. J. Eur. Union, L* **2004**, *23*, 3–16.
30. European Commission. Directive 2002/49/EC of the European parliament and the Council of 25 June 2002 relating to the assessment and management of environmental noise. *Off. J. Eur. Union L* **2002**, *189*, 2002.
31. Ministerio de Agricultura y Pesca, Alimentación y Medio Ambiente. Evaluación de la Calidad del Aire de Espa na 2016. 2017. Available online: https://tinyurl.com/y2jch2qb (accessed on 7 July 2019).
32. *Emerging Issues Update Air Pollution: World'S Worst Environmental Health Risk*; UNEP Year Book; United Nations Environment Programme: Nairobi, Kenya, 2014.
33. European Environment Agency. Air quality in Europe: 2018 Report. European Environment Agency: Copenhagen, Denmark. 2018. Available online: https://www.eea.europa.eu/publications/air-quality-in-europe-2018 (accessed on 7 July 2019),

34. Feng, L.; Zhang, H.; Jiang, Y.; Wang, Z. Evaluation on the Effect of Car Use Restriction Measures in Beijing. In Proceedings of the 51st Annual Transportation Research Forum, Arlington, Virginia, 11–13 March 2010; pp. 1–10.
35. Liu, Y.; Yan, Z.; Liu, S.; Wu, Y.; Gan, Q.; Dong, C. The effect of the driving restriction policy on public health in Beijing. *Nat. Hazards* **2016**, *85*, 751–762. [CrossRef]
36. Wang, L.; Xu, J.; Zheng, X.; Qin, P. Will a Driving Restriction Policy Reduce Car Trips? A Case Study of Beijing, China. In *Environment for Development*; Resources for the Future: Washington, DC, USA, 2013; pp. 1–19.
37. Viard, V.; Fu, S. The effect of Beijing's driving restrictions on pollution and economic activity. *J. Public Econ.* **2015**, *125*, 98–115. [CrossRef]
38. Li, J.; Li, X.B.; Li, B.; Peng, Z.R. The Effect of Nonlocal Vehicle Restriction Policy on Air Quality in Shanghai. *Atmosphere* **2018**, *9*, 299. [CrossRef]
39. Holman, C.; Harrison, R.; Querol, X. Review of the efficacy of low emission zones to improve urban air quality in European cities. *Atmos. Environ.* **2015**, *111*, 161–169. [CrossRef]
40. Panteliadis, P.; Strak, M.; Hoek, G.; Weijers, E.; van der Zee, S.; Dijkema, M. Implementation of a low emission zone and evaluation of effects on air quality by long-term monitoring. *Atmos. Environ.* **2014**, *86*, 113–119. [CrossRef]
41. Boogaard, H.; Janssen, N.A.; Fischer, P.H.; Kos, G.P.; Weijers, E.P.; Cassee, F.R.; van der Zee, S.C.; de Hartog, J.J.; Meliefste, K.; Wang, M.; et al. Impact of low emission zones and local traffic policies on ambient air pollution concentrations. *Sci. Total Environ.* **2012**, *435*, 132–140. [CrossRef]
42. Jensen, S.S.; Ketzel, M.; Nøjgaard, J.K.; Becker, T. What are the impacts on air quality of low emission zones in Denmark. In Proceedings of the Annual Transport Conference at Aalborg University, Aalborg, Denmark, 22 August 2011.
43. Ellison, R.B.; Greaves, S.P.; Hensher, D.A. Five years of London's low emission zone: Effects on vehicle fleet composition and air quality. *Transp. Res. Part D Transp. Environ.* **2013**, *23*, 25–33. [CrossRef]
44. Jones, A.M.; Harrison, R.M.; Barratt, B.; Fuller, G. A large reduction in airborne particle number concentrations at the time of the introduction of "sulphur free" diesel and the London Low Emission Zone. *Atmos. Environ.* **2012**, *50*, 129–138. [CrossRef]
45. Morfeld, P.; Groneberg, D.A.; Spallek, M.F. Effectiveness of low emission zones: Large scale analysis of changes in environmental NO2, NO and NOx concentrations in 17 German cities. *PLoS ONE* **2014**, *9*. [CrossRef]
46. Cyrys, J.; Peters, A.; Soentgen, J.; Wichmann, H.E. Low emission zones reduce PM10 mass concentrations and diesel soot in German cities. *J. Air Waste Manag. Assoc.* **2014**, *64*, 481–487. [CrossRef]
47. Invernizzi, G.; Ruprecht, A.; Mazza, R.; De Marco, C.; Močnik, G.; Sioutas, C.; Westerdahl, D. Measurement of black carbon concentration as an indicator of air quality benefits of traffic restriction policies within the ecopass zone in Milan, Italy. *Atmos. Environ.* **2011**, *45*, 3522–3527. [CrossRef]
48. Ferreira, F.; Gomes, P.; Tente, H.; Carvalho, A.; Pereira, P.; Monjardino, J. Air quality improvements following implementation of Lisbon's Low Emission Zone. *Atmos. Environ.* **2015**, *122*, 373–381. [CrossRef]
49. Santos, F.M.; Gómez-Losada, Á.; Pires, J.C. Impact of the implementation of Lisbon low emission zone on air quality. *J. Hazard. Mater.* **2019**, *365*, 632–641. [CrossRef] [PubMed]
50. da Paz, E.C.; Vieira, T.J.; Zannin, P.H.T. Urban Noise as an Environmental Impact Factor in the Urban Planning Process. In *An Overview of Urban and Regional Planning*; IntechOpen: London, UK, 2018.
51. Han, X.; Huang, X.; Liang, H.; Ma, S.; Gong, J. Analysis of the relationships between environmental noise and urban morphology. *Environ. Pollut.* **2018**, *233*, 755–763. [CrossRef] [PubMed]
52. Labairu, A.; Alsina-Pages, R.M.; Orga, F.; Foraster, M. Noise Annoyance in Urban Life: The Citizen as a Key Point of the Directives. *Proceedings* **2019**, *6*, 5648. [CrossRef]
53. Paneto, G.G.; de Alvarez, C.E.; Zannin, P.H.T. Relationship between urban noise and the health of users of public spaces—A case study in Vitoria, ES, Brazil. *J. Build. Constr. Plan. Res.* **2017**, *5*, 45. [CrossRef]
54. Xu, Y.; Zhu, Y.; Qin, Z. Urban noise mapping with a crowd sensing system. *Wirel. Netw.* **2019**, *25*, 2351–2364. [CrossRef]
55. Recio, A.; Linares, C.; Banegas, J.R.; Díaz, J. Impact of road traffic noise on cause-specific mortality in Madrid (Spain). *Sci. Total Environ.* **2017**, *590*, 171–173. [CrossRef]

56. Amistad, F.T. Assessment of the pedestrianization policy in Vigan City: UNESCO world heritage site. *J. Urban Plan. Dev.* **2010**, *136*, 11–22. [CrossRef]
57. Borge, R.; Lumbreras, J.; Pérez, J.; de la Paz, D.; Vedrenne, M.; de Andrés, J.M.; Rodríguez, M.E. Emission inventories and modeling requirements for the development of air quality plans. Application to Madrid (Spain). *Sci. Total Environ.* **2014**, *466*, 809–819. [CrossRef]
58. Borge, R.; Artíñano, B.; Yagüe, C.; Gomez-Moreno, F.J.; Saiz-Lopez, A.; Sastre, M.; Narros, A.; García-Nieto, D.; Benavent, N.; Maqueda, G.; et al. Application of a short term air quality action plan in Madrid (Spain) under a high-pollution episode-Part I: Diagnostic and analysis from observations. *Sci. Total Environ.* **2018**, *635*, 1561–1573. [CrossRef]
59. Borge, R.; Santiago, J.L.; de la Paz, D.; Martín, F.; Domingo, J.; Valdés, C.; Sánchez, B.; Rivas, E.; Rozas, M.T.; Lázaro, S.; et al. Application of a short term air quality action plan in Madrid (Spain) under a high-pollution episode-Part II: Assessment from multi-scale modelling. *Sci. Total Environ.* **2018**, *635*, 1574–1584. [CrossRef]
60. Toutouh, J.; Lebrusán, I.; Nesmachnow, S. Computational Intelligence for Evaluating the Air Quality in the Center of Madrid, Spain. In Proceedings of the International Conference on Optimization and Learning, Cádiz, Spain, 17–19 February 2020; pp. 115–127.
61. Cohn, L.; McVoy, G. *Environmental Analysis in Transportation Systems.* John Wiley & Sons: Hoboken, NJ, USA, 1982.
62. Robinson, D.W. The concept of noise pollution level. *J. Occup. Environ. Med.* **1971**, *13*, 602. [CrossRef]
63. World Health Organization. Environmental Noise Guidelines for the European Region. 2018. Available online: http://www.euro.who.int/en/health-topics/environment-and-health/noise/publications/2018/environmental-noise-guidelines-for-the-european-region-2018 (accessed on 7 July 2019).
64. Counidad de Madrid. Compendio de Normativa de Ruido y Vibraciones. 2004. Available online: http://www.madrid.org/bdccm/normativa/PDF/Ruidos%20y%20vibraciones/Compilacion/CPRUID.pdf (accessed on 7 July 2019).
65. Chen, W.; Yan, L.; Zhao, H. Seasonal variations of atmospheric pollution and air quality in Beijing. *Atmosphere* **2015**, *6*, 1753–1770. [CrossRef]
66. Palmgren, F.; Berkowicz, R.; Hertel, O.; Vignati, E. Effects of reduction of NOx on the NO2 levels in urban streets. *Sci. Total. Environ.* **1996**, *189–190*, 409 – 415. Highway and Urban Pollution. [CrossRef]
67. Xiao, K.; Wang, Y.; Wu, G.; Fu, B.; Zhu, Y. Spatiotemporal characteristics of air pollutants (PM10, PM2.5, SO2, NO2, O3, and CO) in the inland basin city of Chengdu, southwest China. *Atmosphere* **2018**, *9*, 74. [CrossRef]
68. Thunis, P.; Degraeuwe, B.; Pisoni, E.; Trombetti, M.; Peduzzi, E.; Belis, C.; Wilson, J.; Clappier, A.; Vignati, E. PM2.5 source allocation in European cities: A SHERPA modelling study. *Atmos. Environ.* **2018**, *187*, 93–106. [CrossRef]

Article

Smart Bus Stops as Interconnected Public Spaces for Increasing Social Inclusiveness and Quality of Life of Elder Users

Víctor Manuel Padrón Nápoles [1], Diego Gachet Páez [2,*], José Luis Esteban Penelas [3], Olalla García Pérez [1], María José García Santacruz [4] and Fernando Martín de Pablos [2]

1 Ingeniería Industrial y Aeroespacial, Universidad Europea de Madrid, Villaviciosa de Odón, 28670 Madrid, Spain; victor.padron@universidadeuropea.es (V.M.P.N.); olalla.garpe@gmail.com (O.G.P.)
2 Ciencias y Tecnología de la Información y las Comunicaciones, Universidad Europea de Madrid, Villaviciosa de Odón, 28670 Madrid, Spain; fernando.martin2@universidadeuropea.es
3 Diseño, Arquitectura y Construcciones Civiles, Universidad Europea de Madrid, Villaviciosa de Odón, 28670 Madrid, Spain; jluis.esteban@universidadeuropea.es
4 Facultad de Ciencias Sociales y de la Comunicación, Universidad Europea de Madrid, Villaviciosa de Odón, 28670 Madrid, Spain; mariajose.garcia3@universidadeuropea.es
* Correspondence: diego.gachet@universidadeuropea.es

Received: 6 May 2020; Accepted: 25 May 2020; Published: 1 June 2020

Abstract: As the Smart City concept evolves, it necessarily incorporates more sustainability and inclusiveness features. In this context, the mobility of people is still one of the major challenges for cities. Among the most vulnerable group of citizens are the elderly, as they demand special requirements in the design of smart mobility. At the same time, smart cities' technologies could be used to maintain their quality of life. From an architectural and sociological point of view, smart cities change the meaning and the use of public spaces, from physical meeting places to relational public spaces, in which humans use interposed technological means and information flows. This leads to the concept of Interconnected Public Spaces: a mixture of physical and virtual environments, generating interconnections at a planetary scale, that can be used to attract elderly people for collectively sharing experiences outdoor in public spaces (parks, squares or bus stops, in any city on our planet), increasing their physical form and stimulating them mentally, socially and emotionally. This paper describes the development of an inclusive smart bus stop prototype and the use of its ICT infrastructure to build Interconnected Public Spaces.

Keywords: Smart Cities; inclusive transport; smart bus stops; mobility systems; interconnected public spaces

---

1. Introduction

In the context of Socially Sustainable Cities [1], we certainly have the opportunity to make all city services accessible for everyone. Transportation and mobility systems play a major role in the context of sustainable and inclusive smart cities.

Public transport plays a crucial role for mitigating the social exclusion of vulnerable and disadvantaged groups, affecting their access to basic services and their social and employment relationships. Currently, there are identified specific policies, research priorities and recommendations for local transport, long distance transport and tourism. They address problems such as: (a) the need to combat low awareness of disabled passengers' rights; (b) lack of information on accessibility of local transport; information presented not in accessible formats or not concise and reliable; (c) low use of mobile apps and social media in the sector; (d) low accessibility in suburban and rural areas; (e) general and major access barriers in interchanges and intermodal hubs [2].

In April 2018, the European Commission published the document "Transport in the European Union. Current Trends and Issues", which highlights the importance of social aspects in the development of an advanced European transport system: "From a social perspective, affordability, reliability and accessibility of transport are key. However, this has not been achieved across the board. Addressing these challenges will help pursue sustainable growth in the EU" [3].

The scientific literature in relation to the technological systems used in the transportation systems and their main elements is wide [4]. In addition, there is an important volume of scientific literature describing the use of IoT (Internet of Things) [5,6] technologies for improving the quality of life of citizens in smart cities through measures that lead to a healthy, green and sustainable environment.

Social inclusion is a very complex concept. The EU defines social inclusion as a tendency to enable people at risk of poverty or social exclusion to have the opportunity to participate fully in social life, and thus enjoy an adequate standard of living considered normal in the society in which they live [7]. Social inclusion is especially concerned about people or groups of people who are at risk of deprivation, segregation or marginalization. Special attention deserves the situation of women, as some studies reveal that women have different travel patterns from men and that public transportation plays a crucial role in empowerment, access to opportunities and independence [8].

The definition of social exclusion provided in [9] highlights the importance of mobility in modern cities as a key factor in people's lives, noting that insufficient mobility can cause less accessibility to opportunities, services or social networks.

Therefore, social inclusion and the digitalization of transport have to be harmonized in terms of accessibility, affordability, reliability and inclusiveness, as mentioned in the study about "Social inclusion and EU public transport" [10] for the European Parliament's Committee on Transport and Tourism.

Among the most vulnerable group of citizens are elderly people, they set special requirements in the design of smart mobility, but at the same time, smart cities' technologies could be used to maintain or improve their quality of life.

This paper introduces the novel concept of the Interconnected Public Spaces or IP-Spaces: a mixture of physical and virtual environments, that can be used to attract elderly people for collectively sharing experiences outdoor in public spaces (parks, squares or bus stops, in any city on our planet), in order to increase their physical form and stimulate them mentally, socially and emotionally.

A smart bus stop in development will be the first implementation and validation of the IP-Space concept. The stop can attract elderly users, providing an appropriate interface to a modern digitized transport system and opportunistically use the same ICT infrastructure to provide the IP-Space environment.

The paper is structured as follows. Section 2 introduces premises, the novel concept of IP-Spaces, smart bus stop related work, the architecture of the smart bus stop system being developed and the first approach to IP-Space architecture. Section 3 presents current results of this work in progress. Section 4 analyzes how the current implementation can contribute to improve inclusiveness of elder users in the context of smart mobility, providing specific interfaces and features for this sector of population, including IP-Space activities. Finally, Section 5 describes next steps in the development of IP-Spaces as well as other interfaces for smart bus stops that the authors believe can have a big potential impact on inclusiveness and sustainability in smart mobility.

2. Materials and Methods

As shown in Figure 1, elderly people are one of the most vulnerable group of citizens.

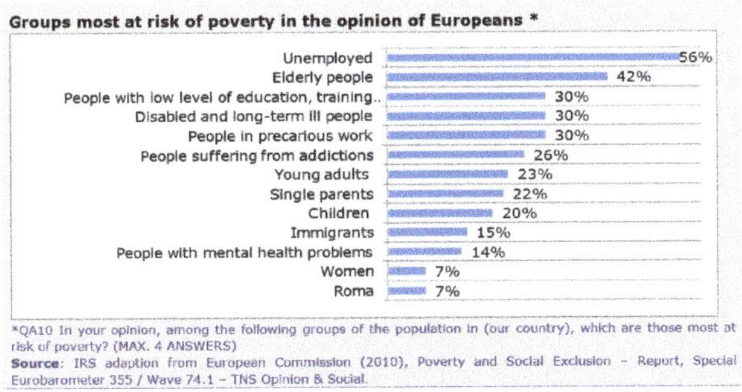

Figure 1. Vulnerable and disadvantaged groups of citizens.

There were 703 million persons aged 65 years or over in the world in 2019. The number of older persons is estimated to double to 1.5 billion in 2050, growing from the current 9% to 16% of the population [11]. This growing sector of the population have special needs to maintain a good physical and mental wellbeing. These needs include physical exercise, social interaction and mental stimulation. Research into ageing and cognition has demonstrated the close relationship of sensory functioning and social communication to maintaining cognitive performance and mood in the elderly, yet in modern societies elderly people are increasingly isolated and under stimulated, both physically and psychosocially [12,13]. This situation results in accelerated cognitive decline and the suffering associated with loneliness and confusion [12]. Social interaction and intellectual stimulation may be relevant to preserving mental functioning in the elderly [14]. Some studies reports that subjects, who participated in senior citizen clubs or senior centers, can lower risk of cognitive decline, especially if this interaction is realized with young adults [15]. Other studies highlight the potential of video games for the development of physical skills, creating mental and social interactions for elderly people, particularly if these video or computer games are designed with an engaging content and provided through an easy and pleasurable interface [16].

2.1. Elder-Oriented Interconnected Public Space

An "Interconnected Public Space" or IP-Space i.e., an *outdoor space* provided with ICT equipment that can connect *only* with a similar space in other part of the world. That means that an IP-Space is a node in a *network of IP-Spaces*. An IP-Space node can be established in a bus stop, in a park or any other outdoor spot in the city, in which a *group of persons* could potentially interact with it. Therefore, IP-Spaces allow the *sharing of collective experiences.*

These nodes can be used to connect with persons from other regions or countries, using the *same or different* languages. They can be used to *participate remotely in different sport, physical, cultural and playful activities* (e.g., interesting games or video games, physical exercises, dancing competitions and many more) *engaging elderly people in remote communities* and stimulating them *physically, socially and intellectually.*

Promoting participation in cultural and sport activities contributes to emotional, physical health and social cohesion [17]. The positive impact of participation in cultural activities—no matter what the level of 'artistic competence' of the people involved—on the perception of one's own psycho-physic wellbeing has been acknowledged for around 40 years and confirmed by a scientific measurement scale, the psychological general wellbeing index. As a conclusion, the connection between culture and subjective wellbeing may often seem obvious although scientific evidence is much harder to get [18]. Interconnected public spaces can help to create the conditions within which wellbeing seems more likely to increase.

As mentioned before, the implementation of the IP-Space can take place at a smart bus stop.

2.2. Smart Bus Stops

The concept of the smart bus stop is recently new as part of the developments related to Smart Cities [19]. Several European cities have launched smart bus stop pilot projects. That is the case of Paris (one stop, Boulevard Diderot, 85 m^2, accessible to persons with disabilities, and providing free Wi-Fi and USB charge, among other services), London (100 Clear Channel bus shelters, using Google Outside service to provide information) and Barcelona (around 10 stops, with mobile-based payment system). Other cities have incorporated some smart elements to traditional stops to supply more information to users, such as arrival time of buses or other general information, without providing more interactivity.

Another example is the smart bus stop prototype of Hungarian company Aquis Innovo in Budapest. The European Commission funded the design and development of the prototype (Figure 2). This prototype, which includes ticket vending, parcel delivery, passenger counting, passenger information, wireless, USB charging, bike rental, air conditioning, taxi ordering, tourist information, news, advertisements, weather forecast, reverse vending, surveillance and other services [20]. On the other hand, there are other smart furniture options, as the outdoor bus ticket-kiosk (Portuguese OEMKIOSK) or information providing smart furniture adapted to people with disabilities such as the Portuguese TOMI as shown in Figure 3.

Figure 2. Aquis Innovo's Smart-Stop (https://europa.eu/investeu/projects/smart-bus-stop_es.EUInvest).

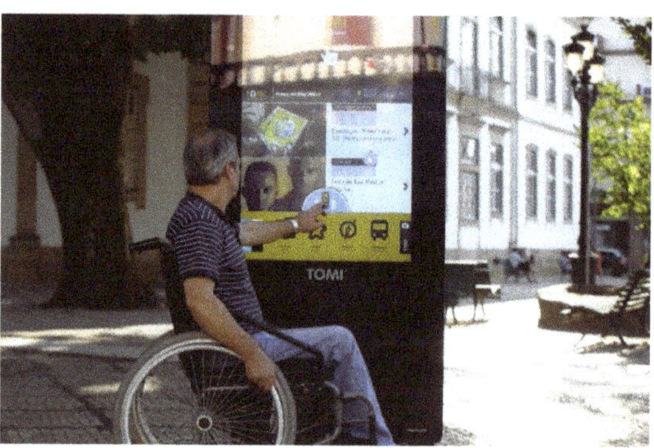

Figure 3. Outdoor TOMI accessible information kiosk (Courtesy TOMI World).

Another example of an advanced pilot project regarding Smart-Stop is the case of Aizuwakamatsu city, Japan. There, the low consumption, bistable e-paper (only consumes power when the message changes) is solar powered and communicated with low power wide area (LPWA) wireless technology to provide information to users. This allows replacing paper timetables and improving the user experience. Managed remotely through the Papercast data management platform, the multilingual displays will present live bus arrivals, timetables, route data, route transfers, service alterations (planned and unplanned) and a range of other travel advices [21].

Despite the huge potential of smart bus stops, their penetration in many European cities is very limited and its adaptation to inclusiveness is just starting to be developed.

2.3. MUSA Smart Bus Stop

MUSA (Advanced Sustainable Urban Furniture–Mobiliario Urbano Sostenible y Avanzado) is the smart bus stop under development in the city of Madrid that will also host an IP-Space. Its main characteristic is the provision of information services with a focus on inclusive and socially-driven transport aspects.

The smart bus stop, from the inclusiveness point of view, is:

(a) An interactive bus stop available to the whole population. It is a public access point to a digitized transport system (DTS), which allows access to persons without apps or even without a smart phone.
(b) It can work as a public access point and as a travel assistant for low-income or disadvantaged groups of users.
(c) It can improve the accessibility to DTS through customization of interfaces and reduction of cognitive demand.
(d) It can improve planning in real-time taking into account unexpected events that can improve or disrupt transport operations.
(e) Through an attractive and customized interface, it can foster the penetration of travel planning apps and its use by different user groups (elders, immigrants, etc.).
(f) It can be implemented as a reduced size smart furniture providing a robust, essential electronic equipment that converts traditional stops into accessible smart bus stops, minimizing the modernization cost and having a wide use in the cities and rural areas.
(g) Finally, it can be used for introducing Interconnected Public Space spots of the city. This is very suitable when the smart bus stop is in a park or square, where people can be involved in sport, physical and cultural activities. Sharing the same ICT infrastructure makes the system attractive from an aesthetical and economic point of view.

2.3.1. MUSA Smart Bus Stop System Architecture

MUSA is a physical stop equipped with an interactive display (currently a 40-inch monitor provided with an infrared frame) and a computer system communicating with a set of cloud systems to provide different services to travelers at the stop. These services available in the stop are called Group Services (Figure 4).

The main point of these services is a multimodal travel planner including options for walking, cycling, private and public transport. The interface to this urban equipment is being customized to increase the accessibility for all citizens and particularly for those vulnerable to exclusion.

In addition, a community service is available for planning and realization of IP-Space related activities. Services related to publicity, environment and health are also planned, without excluding others that can be included in the future.

Transport services and applications provided in the stop can also be available for mobile phones. These apps are called Individual Services (Figure 4). Some of these apps are currently implemented using web technologies and accessing third parties' services through their APIs. Other apps, still under development, will use our own developed cloud-based services. Through the use of Individual Services, the smart bus stop can help to foster transport planning in general, not only when passengers are waiting in the stop, but also in any other places using mobile phones or other devices. A massive planning of transport needs can help to develop new ways of organizing transport [22].

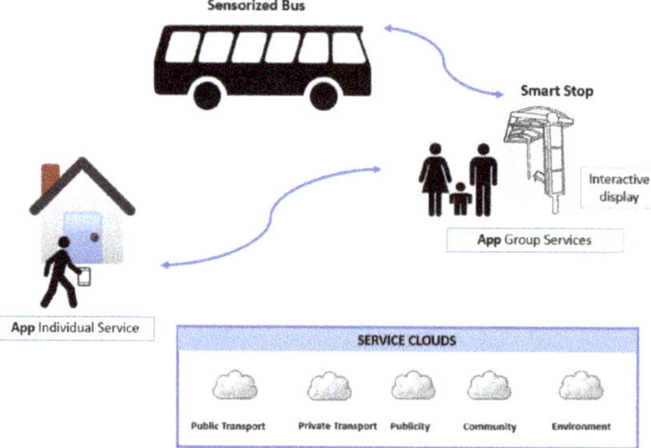

Figure 4. Main components of MUSA system architecture.

The planning of users' transport needs helps to characterize the demand on transport systems. This identified demand jointly with sensorization of transport means can be used by transport providers to customize and fine-tune their services in order to meet user requirements (Figure 5) and increase satisfaction of their clients.

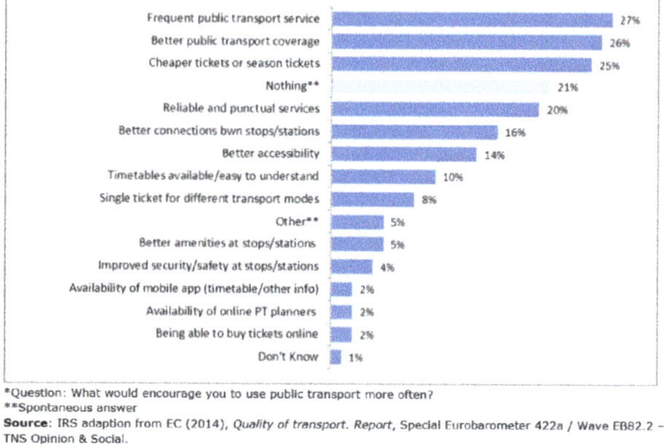

Figure 5. Factors that can encourage the use of public transport (Roma refers to Romani people).

A key factor in the users' transport planning is to know or estimate real-time bus occupancy. Otherwise, this planning can be useless. This factor is taken into account in MUSA system.

2.3.2. Sensorization of Buses

The efficiency of planning, from users' as well as from transport providers' point of view, is highly correlated to the level of sensorization of transport means. In the context of MUSA project, this led to increased sensorization of buses installing Automatic Passenger Counters (APC) to know the occupancy of the bus in real-time, the availability of free places for wheelchairs and baby strollers, as well as the flow of passengers in each bus stop [22,23].

The use of APC for transport providers is very important to analyze the performance of bus services in real-time, to detect the most demanded routes or segment of routes and potentially replan them to increase efficiency of service and users' satisfaction. From users' point of view, the bus occupancy (or its probability) is crucial for effective planning (it is useless, if the planned bus comes full and passengers cannot get on board).

Buses normally include AVL (Automatic Vehicle Location) using GPS and SC-AFC (Smart Card Automatic Fare). For a flat rate service (where passengers do not check out when alighting the bus, as in Madrid), the next level of sensorization was the use of APC or Automatic Passenger Counters. There are different technologies for APC, for example, infrared systems and vision systems (video cameras, stereo cameras and time-of-flight cameras).

The use of the latter can be adapted to detect free available places for wheelchairs and baby strollers, thereby supporting greater inclusiveness.

Though for this project, infrared systems and stereo cameras were studied, MUSA used simple video CCTV cameras from Retail Sensing, a Manchester company, and evaluated their performance in real-life conditions (Figure 6). The cameras, located on top of the front and rear doors, use artificial vision algorithms to count in and out passengers. This information was sent through a 4G router to an MQTT server to make it globally available.

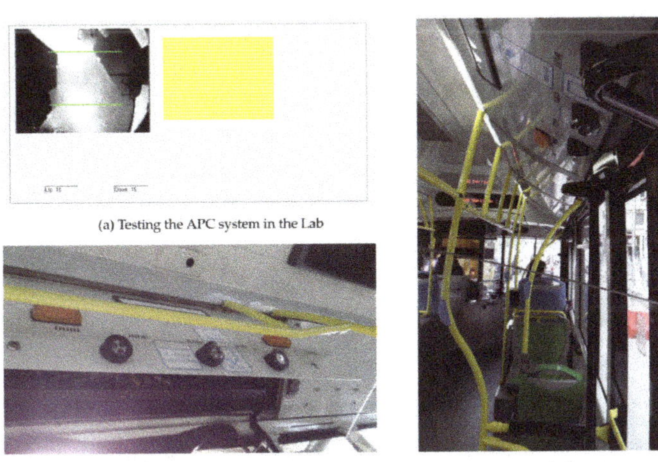

(a) Testing the APC system in the Lab

(b) Installing the APC system in the bus

(c) Testing during daily operations

Figure 6. Installing and testing the Automatic Passenger Counters (APC) based on video cameras.

First, we tested the camera system in the Lab; next, we installed it inside a bus and tested it during daily operations in the center of Madrid (Figures 7 and 8).

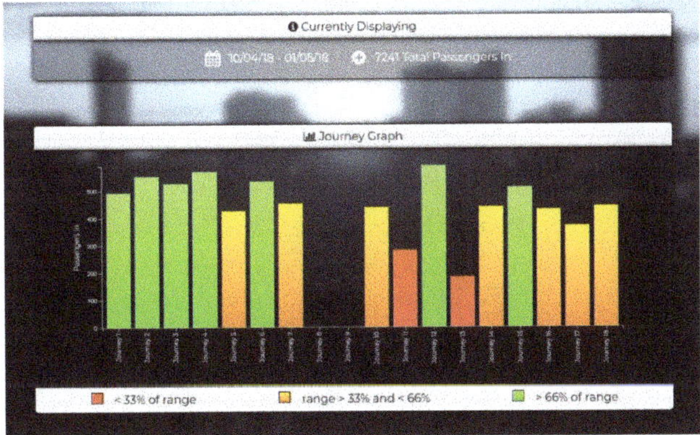

Figure 7. Number of daily passengers entering the bus (Courtesy Retail Sensing).

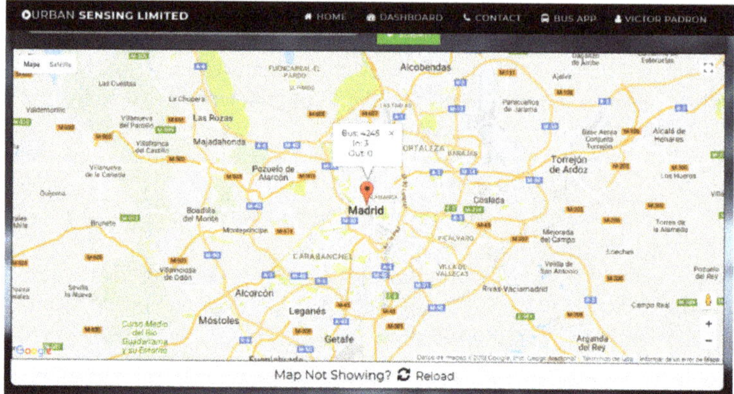

Figure 8. Behavior of passengers' flow at one bus stop (Courtesy Retail Sensing).

Currently, a massive installation of APC systems using time-of-flight and Artificial Intelligence systems is taking place on Madrid buses. It is foreseeable that information about real-time occupancy of buses and availability of wheelchairs and baby strollers will be publicly available as Open Data.

2.3.3. MUSA Transport Services

The multimodal trip planner employed in MUSA includes options for walking, cycling, private and public transport. It is designed as a special software layer (Figure 9) that can run on a commercial travel planner, such as Google Maps (and its API). This approach has three advantages: it allows the customization of interfaces for different users' segments; as elderly people, it allows the anonymous collection of traveling data for building mobility models and developing social innovation solutions (Social Model in Figure 9); and finally, it can be adapted to different commercial planners.

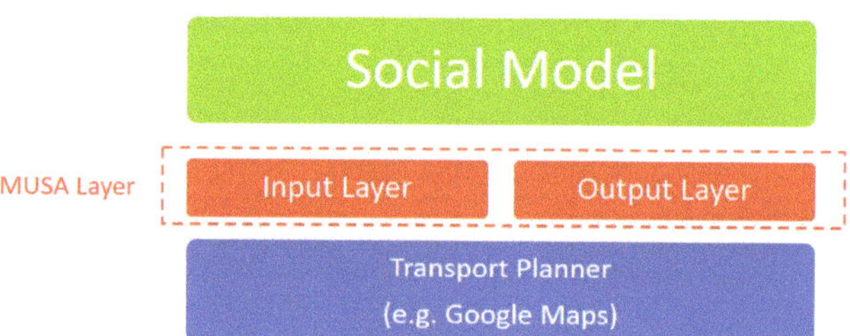

Figure 9. Layering of the multimodal planning application.

2.3.4. Interconnected Public Space Service in MUSA

The new functionality of the Interconnected Public Space (IP-Space) is being developed in the current prototype of a MUSA smart bus stop.

From the stop, which is an IP-Space node, a user or set of users can connect to persons from other regions or countries present in another IP-Space node, using the same or different languages. All can participate remotely in different sport, physical, cultural and playful activities (e.g., interesting games or video games, physical exercises, dancing competitions and many more). The interface can transmit audio or video using cameras and microphones, although this can be restricted in accordance with current local privacy laws. This also will depend on whether the IP-Space is physically enclosed or if it is a completely open public location. However, in all cases, the interface will allow the transmission of images, graphics and text as basic means of communication.

In a first approach to the design of the prototype, the high-level operation of this service consists of three steps (Figure 10):

(a) Map. In this step, users request the realization of a given activity to other people in a remote IP-Space. They can agree to start the activity immediately or schedule it at a given date and time.
(b) Menu. They select the desired or scheduled activity from the repository and access it.
(c) Activity. They proceed to realize the desired activity. They will have a language assistance application, so people from different countries and culture can communicate.

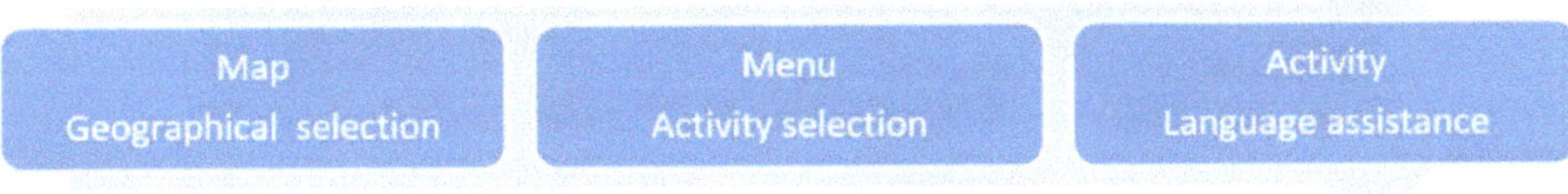

Figure 10. IP-Space users' interaction sequence.

3. Results

The current state of the prototype of the MUSA smart bus stop and its main interface are shown in Figure 11. An advertisement is running in the background (in this case, for musicals in the center of Madrid), while different services are available in the lower carrousel. For intuitively attracting users to different interfaces, two type of icons are available. A traditional picture is used to attract more resolute and direct users (e.g., some senior people). Let us call it the "conservative interface". Moreover, a more playful icon is used to attract more skillful and playful users (e.g., some junior or young people). Let us call this the "playful interface". Other future special services for assisting travelers with special needs (elderly, those with reduced mobility, easier travel with kids or pregnant women) are also included in the interface.

Figure 11. Main interface of the smart bus stop prototype.

The conservative interface of the public transport app is shown in Figure 12. A box with the most frequently used destination from the current stop is shown in the top right corner. This helps to increase the probabilities of reducing the interaction to minimum. Below, a box shows time and duration of the selected trip. In addition, there is an option to select private or other type of alternative transport without leaving the conservative mode of interaction.

Figure 12. Multimodal planner using public transport.

A third box allows users to select any origin and destination, using a tactile keyboard on the screen. Destinations are dynamically ordered based on their frequency of use. That means that the stop system can learn from users' interactions the required destinations at each time of the day during the week.

Finally, information about nearest bus stops can be searched. Typical interactive features of Google Maps are disabled, so conservative users cannot be distracted from their simple, direct interaction with the app.

Figure 13 shows the conservative interface for fostering physical exercises by means of walking. In this case, time and duration of a trip walking and using public transport are similar, so the option of walking can be healthier for the user. This feature can be very interesting for elderly people.

Figure 13. Fostering physical exercise.

Another interesting service for elderly people (under development) is the possibility to use mobility service providers as Uber, Cabify or Free Now, without having the application or even having a smartphone (Figure 14). The private transport interface of the MUSA smart bus stop can check the availability of these services and order a trip using a dedicated account. The user will pay using cash, as this payment option is currently available in big cities.

The main interface design of the IP-Space is shown in Figure 15. Again, there are two icons for conservative and playful interfaces, labelled as "Community", which give access to IP-Space services. A third icon is reserved for future services to the elderly.

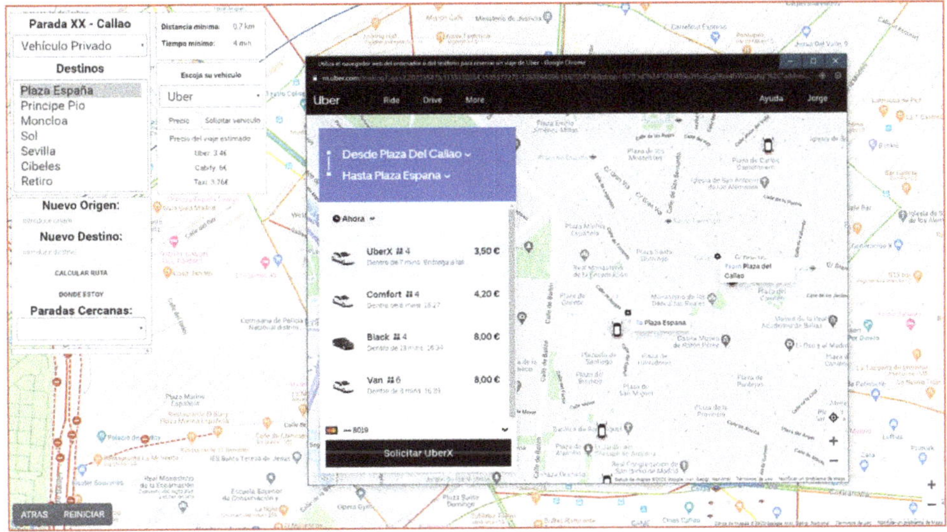

Figure 14. Ordering a mobility service from the smart bus stop.

Figure 15. Interconnected public space access interface.

4. Discussion

This work in progress is an experimental platform, which has two main objectives: increasing social inclusiveness and quality of life of the elderly.

First, it addresses the inclusiveness in general, and especially of elderly people in smart transport systems (digitized transport systems). For the elderly, a customization of interfaces is proposed to reduce cognitive demand in the use of a multimodal transport planner application. The use of an easy interface can foster that users employ the same apps available in the MUSA stop (Group Services) in their mobile phones (Individual Services).

Sensorizing buses to detect or estimate buses' real-time occupancy is crucial for increasing the effectiveness of transport planning, from both users' and transport providers' points of view. The use of sensors as time-of-flight cameras for this task can be used to detect or estimate vacant places for wheelchairs and baby strollers, therefore increasing inclusiveness.

Therefore, this work is addressing several issues related to inclusiveness as the lack of information on accessibility of local transport, the lack of information presented in accessible formats or concise and reliable, and the low use of mobile apps and social media in the local transport sector.

Increasing the use of planning apps have the potential to anonymously create mathematical models for planning mobility at city scale and therefore, opens the possibility of new transportation paradigms. In addition, socialization at smart bus stops allows citizens to potentially organize and require new transport options.

Other features providing for elderly users are the fostering of walking (physical exercise) and the chance to friendly order mobility services as Uber, Cabify, FreeNow, etc., increasing their trust on these new transport options.

Finally, this paper introduces the novel concept of Interconnected Public Space or IP-Space and describe its potential for increasing (or at least maintain) quality of life of the elderly. Basic design of IP-Space is described, as well as how to link it with smart bus stop development, to take advantage of ICT smart bus stop infrastructure. This can be particularly effective in wide outdoor spaces as parks and squares, producing a faster and cheaper deployment of IP-Spaces.

5. Conclusions

Interconnected Public Spaces are in a very early stage of development and sociological research; there are several lines of research to explore as soon as the prototype will be ready. It will need to address sociological and psychological acceptance, legal issues in different countries, the development of the activities, the physical construction of the IP-Spaces from an architectural point of view, etc.

Regarding the smart bus stop, the next step in research is the study of more advanced interfaces, as those provided by the Voice Activated Personal Assistants or VAPA (e.g., Alexa), taking into account privacy concerns [24] and cybersecurity. This can require the design of a physical interface in the smart furniture, creating a micro space that isolates users and makes easier voice interaction. This type of interface can also help to foster the use of VAPAs at home. Furthermore, VAPA systems have the potential to increase accessibility of disabled people in general [25] and the elderly [26], as well as decrease the cognitive load and the effort of planning travels in the context of modern digitized transport systems. Finally, there are technical and research challenges related to the performance of VAPA systems outdoors with multiple voices and the improving of voice interface for elderly people [27].

Author Contributions: Conceptualization and Methodology, V.M.P.N., D.G.P., O.G.P. and J.L.E.P.; Software design, V.M.P.N., M.J.G.S. and O.G.P.; Investigation, M.J.G.S., D.G.P., J.L.E.P., V.M.P.N., O.G.P. and F.M.d.P.; writing—review and editing, V.M.P.N., D.G.P. and J.L.E.P. All authors have read and agreed to the published version of the manuscript.

Funding: This research was funded by Universidad Europea de Madrid by grants numbered 2017/UEM22, 2018/UEM06 and 2019/UEM19.

Acknowledgments: We want to express our deep gratitude to all those persons and institutions that are helping and supporting us in the realization of this project. We want to thank the institutional support from the Municipal Transport Company (EMT) of Madrid, for the kind support of its executives and technical staff; Asad Syed from Retail Sensing for his kind support and sponsorship with the APC system; and our students, Alba Gutiérrez García-Ochoa, Alfonso López Pérez, Germán García García and Jorge García González. To all of you, thank you very much.

Conflicts of Interest: The authors declare no conflict of interest. All trademarks are the property of their respective owners.

References

1. Verebes, T. (Ed.) *Masterplanning the Adaptive City: Computational Urbanism in the Twenty-First Century*, 1st ed.; Routledge, Taylor and Francis Group: London, UK; New York, NY, USA, 2014.
2. Bekiaris, E.; Loukea, M.; Spanidis, P.; Ewing, S.; Denninghaus, M.; Ambrose, I.; Papamichail, K.; Castiglioni, R.; Veitch, C. *Research for TRAN Committee: Transport and Tourism for Persons with Disabilities and Persons with Reduced Mobility*; European Parliament, Policy Department for Structural and Cohesion Policies: Brussels, Belgium, 2018; p. 19.
3. Transport in the European Union: Current Trends and Issues. Available online: https://ec.europa.eu/transport/sites/transport/files/2018-transport-in-the-eu-current-trends-and-issues.pdf (accessed on 25 April 2020).

4. Kehua, S.; Jie, L.; Hongbo, F. Smart city and the applications. In Proceedings of the International Conference on Electronics, Communications and Control (ICECC), Ningbo, China, 9–11 September 2011; pp. 1028–1031.
5. Mohanty, S.P.; Choppali, U.; Kougianos, E. Everything you wanted to know about smart cities: The Internet of things is the backbone. *IEEE Consum. Electron. Mag.* **2016**, *5*, 60–70. [CrossRef]
6. Cisco España. Internet of Things World Forum Barcelona 2013 (Video, November 2013). Wallbank, P. Touring the Barcelona Smart City Project. Available online: http://paulwallbank.com/2014/06/09/geeks-tour-of-smartcity-barcelona-slideshow-cisco-iot/ (accessed on 27 March 2020).
7. Poverty and Social Exclusion. Available online: https://ec.europa.eu/social/main.jsp?catId=751&langId=en (accessed on 24 April 2020).
8. The Role of Women in the Green Economy. The Issue of Mobility. Available online: http://www.europarl.europa.eu/RegData/etudes/note/join/2012/462453/IPOL-FEMM_NT(2012)462453_EN.pdf (accessed on 24 April 2020).
9. Kenyon, K.; Lyons, G.; Rafferty, J. Transport and social exclusion: Investigating the possibility of promoting social inclusion through virtual mobility. *J. Transp. Geogr.* **2003**, *10*, 207–219. [CrossRef]
10. Social Inclusion in EU Public Transport. Available online: https://www.europarl.europa.eu/RegData/etudes/STUD/2015/540351/IPOL_STU(2015)540351_EN.pdf (accessed on 24 April 2020).
11. Department of Economic and Social Affairs, United Nations. *World Population Ageing 2019: Highlights*; United Nations: New York, NY, USA, 2019; pp. 1–2.
12. Waterworth, J.; Ballesteros, S.; Christian, P.; Bieber, G.; Kreiner, A.; Wiratanaya, A.; Polymenakos, L.; Wanche-Politis, S.; Capobianco, M.; Etxeberria, I.; et al. Ageing in a networked society—Social inclusion and mental stimulation. In Proceedings of the 2nd International Conference on Pervasive Technologies Related to Assistive Environments (PETRA 2009), Corfu, Greece, 9–13 June 2009.
13. Rabbitt, P. Cognitive changes across the lifespan. In *The Cambridge Handbook of Age and Ageing*, 1st ed.; Johnose, M.L., Bengston, V.L., Coleman, P.G., Kirkwood, T.B.L., Eds.; Cambridge University Press: Cambridge, UK, 2005; pp. 190–199.
14. Wang, H.; Karp, A.; Winblad, B.; Fratiglioni, L. Late-Life Engagement in Social and Leisure Activities Is Associated with a Decreased Risk of Dementia: A Longitudinal Study from the Kungsholmen Project. *Am. J. Epidemiol.* **2002**, *155*, 1081–1087. [CrossRef] [PubMed]
15. Lee, S.H.; Kim, Y.B. Which type of social activities may reduce cognitive decline in the elderly?: A longitudinal population-based study. *BMC Geriatr.* **2016**, *16*, 165. [CrossRef] [PubMed]
16. IJsselsteijn, W.; Nap, H.H.; de Kort, Y.; Poels, K. Digital Game Design for Elderly Users. In Proceedings of the Conference on Future Play—Future Play '07, Toronto, ON, Canada, 15–17 November 2007.
17. Brophy, M.; Mguni, N.; Mulgan, G.; Shandro, A. *The State of Happiness: Can Public Policy Shape People's Wellbeing and Resilience?* The Young Foundation: London, UK, 2010; Available online: https://youngfoundation.org/wp-content/uploads/2012/10/The-State-of-Happiness.pdf (accessed on 22 April 2020).
18. Diener, E. *The Science of Well-Being: The Collected Works of Ed Diener*; Social Indicators Research Series; Springer: London, UK, 2009; Volume 37.
19. Gretzel, U.; Sigala, M.; Xiang, Z.; Koo, C. Smart tourism: Foundations and developments. *Electron. Mark.* **2015**, *25*, 179–188. [CrossRef]
20. Aquis Innovo. Available online: https://europa.eu/investeu/projects/smart-bus-stop_es (accessed on 22 April 2020).
21. Papercast 2018. Available online: https://www.papercast.com/epaper_bus_stop_passenger_information_solutions/products_e-paper_displays/ (accessed on 22 April 2020).
22. Nápoles, V.M.P.; Rodríguez, M.B.; Páez, D.G.; Penelas, J.L.E.; García-Ochoa, A.G.; Pérez, A.L. MUSA–I. Towards New Social Tools for Advanced Multi-Modal Transportation in Smart Cities. *Proceedings* **2018**, *2*, 1215.
23. Nápoles, V.M.P.; Páez, D.G.; Penelas, J.L.E.; García, G.G.; Santacruz, M.J.G. Bus Stops as a Tool for Increasing Social Inclusiveness in Smart Cities. In *Smart Cities. ICSC-CITIES 2019. Communications in Computer and Information Science*, 1st ed.; Nesmachnow, S., Hernández Callejo, L., Eds.; Springer: Cham, Switzerland, 2020; Volume 1152, pp. 215–227.
24. Moorthy, A.E.; Vu, K.L.V. Privacy Concerns for Use of Voice Activated Personal Assistant in the Public Space. *Int. J. Hum. Comput. Interact.* **2015**, *31*, 307–335. [CrossRef]
25. Pradhan, A.; Mehta, K.; Findlater, L. "Accessibility Came by Accident": Use of Voice-Controlled Intelligent Personal Assistants by People with Disabilities. In Proceedings of the 2018 CHI Conference on Human Factors in Computing Systems, Montreal, QC, Canada, 21–26 April 2018; pp. 1–13.

26. O'Brien, K.; Liggett, A.; Ramirez-Zohfeld, V.; Sunkara, P.; Lindquist, L.A. Voice-Controlled Intelligent Personal Assistants to Support Aging in Place. *J. Am. Geriatr. Soc.* **2020**, *68*, 176–179.
27. Sayago, S.; Neves, B.B.; Cowan, B.R. Voice assistants and older people: Some open issues. In Proceedings of the 1st International Conference on Conversational User Interfaces, Dublin, Ireland, 22–23 August 2019; pp. 1–3.

MDPI
St. Alban-Anlage 66
4052 Basel
Switzerland
Tel. +41 61 683 77 34
Fax +41 61 302 89 18
www.mdpi.com

Smart Cities Editorial Office
E-mail: smartcities@mdpi.com
www.mdpi.com/journal/smartcities

www.ingramcontent.com/pod-product-compliance
Lightning Source LLC
LaVergne TN
LVHW070619170726
843515LV00004B/130
* 9 7 8 3 0 3 9 4 3 0 5 0 5 *